W9-ACE-994

A SHORT HISTORY OF THE WORLD

¶ Mr. WELLS has also written the following novels:

LOVE AND MR. LEWISHAM
KIPPS
MR. POLLY
THE WHEELS OF CHANCE
THE NEW MACHIAVELLI
ANN VERONICA
TONO BUNGAY
MARRIAGE
BEALBY
THE PASSIONATE FRIENDS
THE WIFE OF SIR ISAAC HARMON
THE RESEARCH MAGNIFICENT
MR. BRITLING SEES IT THROUGH
THE SOUL OF A BISHOP
JOAN AND PETER
THE UNDYING FIRE

¶ The following fantastic and imaginative romances:

THE WAR OF THE WORLDS
THE TIME MACHINE
THE WONDERFUL VISIT
THE ISLAND OF DR. MOREAU
THE SEA LADY
THE SLEEPER AWAKES
THE FOOD OF THE GODS
THE WAR IN THE AIR
THE FIRST MEN IN THE MOON
IN THE DAYS OF THE COMET
THE WORLD SET FREE

And numerous Short Stories now collected in One Volume under the title of
THE COUNTRY OF THE BLIND

¶ A Series of books on Social, Religious, and Political questions:

ANTICIPATIONS (1900)
MANKIND IN THE MAKING
FIRST AND LAST THINGS
NEW WORLDS FOR OLD
A MODERN UTOPIA
THE FUTURE IN AMERICA
AN ENGLISHMAN LOOKS AT THE WORLD
WHAT IS COMING?
WAR AND THE FUTURE
IN THE FOURTH YEAR
GOD THE INVISIBLE KING
THE OUTLINE OF HISTORY
WASHINGTON AND THE RIDDLE OF PEACE

¶ And two little books about children's play, called

FLOOR GAMES and LITTLE WARS

A Short History of The World

BY

H. G. WELLS

ILLUSTRATED

THE REVIEW OF REVIEWS CORPORATION

PUBLISHERS NEW YORK

1923

Published by arrangement with the Macmillan Company

PRINTED IN THE UNITED STATES OF AMERICA

Set up and electrotyped. Published October, 1922.

PREFACE

THIS SHORT HISTORY OF THE WORLD is meant to be read straightforwardly almost as a novel is read. It gives in the most general way an account of our present knowledge of history, shorn of elaborations and complications. It has been amply illustrated and everything has been done to make it vivid and clear. From it the reader should be able to get that general view of history which is so necessary a framework for the study of a particular period or the history of a particular country. It may be found useful as a preparatory excursion before the reading of the author's much fuller and more explicit *Outline of History* is undertaken. But its especial end is to meet the needs of the busy general reader, too driven to study the maps and time charts of that *Outline* in detail, who wishes to refresh and repair his faded or fragmentary conceptions of the great adventure of mankind. It is not an abstract or condensation of that former work. Within its aim the *Outline* admits of no further condensation. This is a much more generalized History, planned and written afresh.

H. G. WELLS.

CONTENTS

Contents

LIST OF ILLUSTRATIONS

List of Illustrations

List of Illustrations

A SHORT HISTORY OF THE WORLD

A SHORT HISTORY OF THE WORLD

I

The World in Space

THE story of our world is a story that is still very imperfectly known. A couple of hundred years ago men possessed the history of little more than the last three thousand years. What happened before that time was a matter of legend and speculation. Over a large part of the civilized world it was believed and taught that the world had been created suddenly in 4004 B.C., though authorities differed as to whether this had occurred in the spring or autumn of that year. This fantastically precise misconception was based upon a too literal interpretation of the Hebrew Bible, and upon rather arbitrary theological assumptions connected therewith. Such ideas have long since been abandoned by religious teachers, and it is universally recognized that the universe in which we live has to all appearances existed for an enormous period of time and possibly for endless time. Of course there may be deception in these appearances, as a room may be made to seem endless by putting mirrors facing each other at either end. But that the universe in which we live has existed only for six or seven thousand years may be regarded as an altogether exploded idea.

The earth, as everybody knows nowadays, is a spheroid, a sphere slightly compressed, orange fashion, with a diameter of nearly 8,000 miles. Its spherical shape has been known at least to a limited number of intelligent people for nearly 2,500 years, but before that time it was supposed to be flat, and various ideas which now seem fantastic were entertained about its relations to the sky and the stars and planets. We know now that it rotates upon its axis (which is about 24 miles shorter than its equatorial diameter) every twenty-four hours, and that this is the cause of the alternations of day and night, that it circles about the sun in a slightly distorted and slowly variable oval path in a year. Its distance from the sun varies between ninety-one and a half millions at its nearest and ninety-four and a half million miles.

About the earth circles a smaller sphere, the moon, at an average distance of 239,000 miles. Earth and moon are not the only bodies to travel round the sun. There are also the planets, Mercury and Venus, at distances of thirty-six and sixty-seven millions of miles; and beyond the circle of the earth and disregarding a belt of numerous smaller bodies, the planetoids, there are Mars, Jupiter, Saturn, Uranus and Neptune at mean distances of 141, 483, 886, 1,782, and 1,793 millions of miles respectively. These figures in

Photo: G. W. Ritchey

"LUMINOUS SPIRAL CLOUDS OF MATTER"

(*Nebula photographed* 1910)

Photo: G. W. Ritchey

THE NEBULA SEEN EDGE-ON

Note the central core which, through millions of years, is cooling to solidity

millions of miles are very difficult for the mind to grasp. It may help the reader's imagination if we reduce the sun and planets to a smaller, more conceivable scale.

If, then, we represent our earth as a little ball of one inch diameter, the sun would be a big globe nine feet across and 323 yards away, that is about a fifth of a mile, four or five minutes' walking. The moon would be a small pea two feet and a half from the world. Between earth and sun there would be the two inner planets, Mercury and Venus, at distances of one hundred and twenty-five and two hundred and fifty yards from the sun. All round and about these bodies there would be emptiness until you came to Mars, a hundred and seventy-five feet beyond the earth; Jupiter nearly a mile away, a foot in diameter; Saturn, a little smaller, two miles off; Uranus four miles off and Neptune six miles off. Then nothingness and nothingness except for small particles and drifting scraps of attenuated vapour for thousands of miles. The nearest star to earth on this scale would be 40,000 miles away.

These figures will serve perhaps to give one some conception of the immense emptiness of space in which the drama of life goes on.

For in all this enormous vacancy of space we know certainly of life only upon the surface of our earth. It does not penetrate much more than three miles down into the 4,000 miles that separate us from the centre of our globe, and it does not reach more than five miles above its surface. Apparently all the limitlessness of space is otherwise empty and dead.

The deepest ocean dredgings go down to five miles. The highest recorded flight of an aeroplane is little more than four miles. Men have reached to seven miles up in balloons, but at a cost of great suffering. No bird can fly so high as five miles, and small birds and insects which have been carried up by aeroplanes drop off insensible far below that level.

II

The World in Time

IN the last fifty years there has been much very fine and interesting speculation on the part of scientific men upon the age and origin of our earth. Here we cannot pretend to give even a summary of such speculations because they involve the most subtle mathematical and physical considerations. The truth is that the physical and astronomical sciences are still too undeveloped as yet to make anything of the sort more than an illustrative guesswork. The general tendency has been to make the estimated age of our globe longer and longer. It now seems probable that the earth has had an independent existence as a spinning planet flying round and round the sun for a longer period than 2,000,000,000 years. It may have been much longer than that. This is a length of time that absolutely overpowers the imagination.

Before that vast period of separate existence, the sun and earth and the other planets that circulate round the sun may have been a great swirl of diffused matter in space. The telescope reveals to us in various parts of the heavens luminous spiral clouds of matter, the spiral nebulæ, which appear to be in rotation about a centre. It is supposed by many astronomers that the sun and its planets were once such a spiral, and that their matter has undergone concentration into its present form. Through majestic æons that concentration went on until in that vast remoteness of the past for which we have given figures, the world and its moon were distinguishable. They were spinning then much faster than they are spinning now; they were at a lesser distance from the sun; they travelled round it very much faster, and they were probably incandescent or molten at the surface. The sun itself was a much greater blaze in the heavens.

If we could go back through that infinitude of time and see the earth in this earlier stage of its history, we should behold a scene more like the interior of a blast furnace or the surface of a lava flow before it cools and cakes over than any other contemporary scene.

Photo: G. W. Ritchey

THE GREAT SPIRAL NEBULA

No water would be visible because all the water there was would still be superheated steam in a stormy atmosphere of sulphurous and metallic vapours. Beneath this would swirl and boil an ocean of molten rock substance. Across a sky of fiery clouds the glare of the hurrying sun and moon would sweep swiftly like hot breaths of flame.

Slowly by degrees as one million of years followed another, this

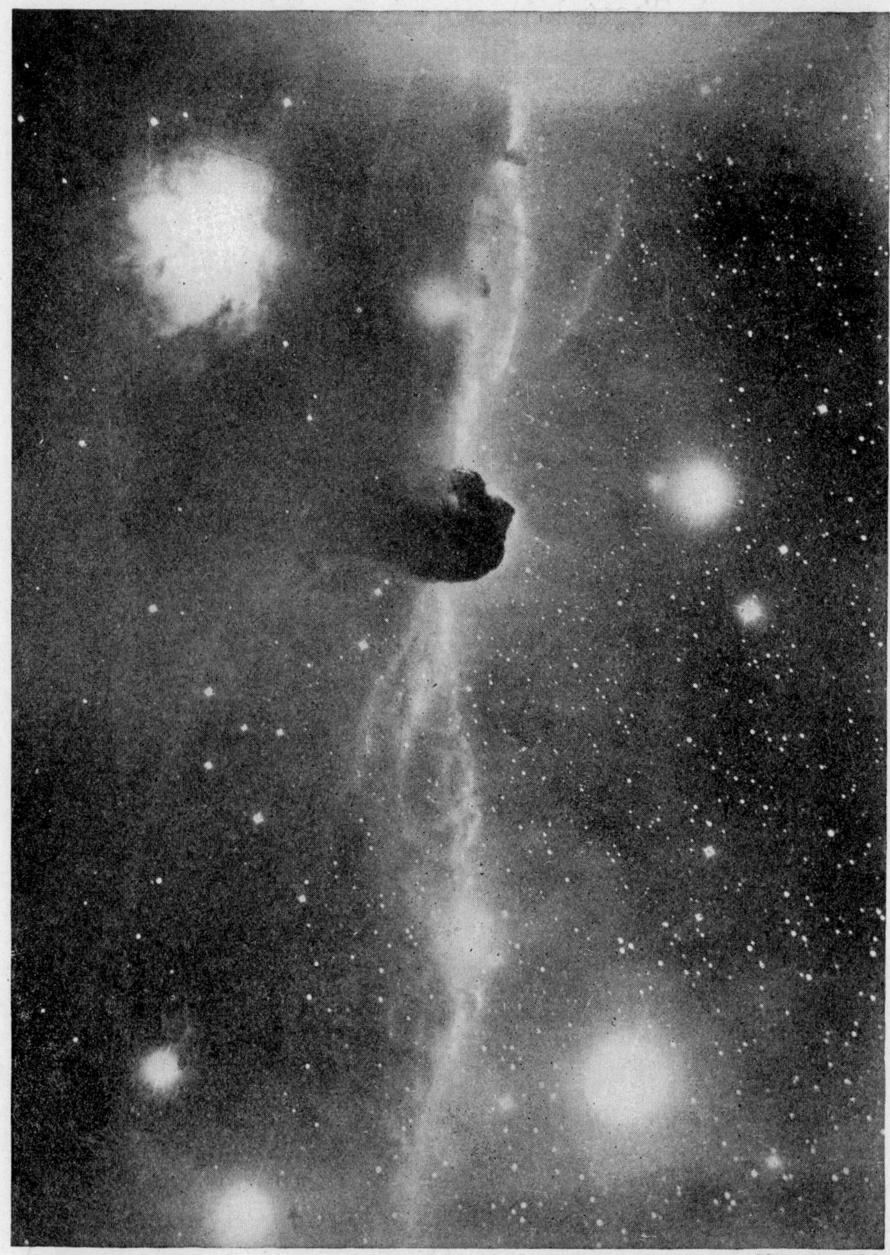

Photo: Prof. Hale

A DARK NEBULA

Taken in 1920 with the aid of the largest telescope in the world. One of the first photographs taken by the Mount Wilson telescope

There are dark nebulæ and bright nebulæ. Prof. Henry Norris Russell, against the British theory, holds that the dark nebulæ preceded the bright nebulæ

fiery scene would lose its eruptive incandescence. The vapours in the sky would rain down and become less dense overhead; great slaggy cakes of solidifying rock would appear upon the surface of the molten sea, and sink under it, to be replaced by other floating masses. The sun and moon growing now each more distant and each smaller, would rush with diminishing swiftness across the heavens. The moon now, because of its smaller size, would be already cooled far below incandescence, and would be alternately obstructing and reflecting the sunlight in a series of eclipses and full moons.

And so with a tremendous slowness through the vastness of time, the earth would grow more and more like the earth on which we live, until at last an age would come when, in the cooling air, steam would begin to condense into clouds, and the first rain would fall hissing upon the first rocks below. For endless millenia the greater part of the earth's water would still be vaporized in the atmosphere, but there would now be hot streams running over the crystallizing

Photo: G. W. Ritchey

ANOTHER SPIRAL NEBULA

LANDSCAPE BEFORE LIFE
"Great lava-like masses of rock without traces of soil"

rocks below and pools and lakes into which these streams would be carrying detritus and depositing sediment.

At last a condition of things must have been attained in which a man might have stood up on earth and looked about him and lived. If we could have visited the earth at that time we should have stood on great lava-like masses of rock without a trace of soil or touch of living vegetation, under a storm-rent sky. Hot and violent winds, exceeding the fiercest tornado that ever blows, and downpours of rain such as our milder, slower earth to-day knows nothing of, might have assailed us. The water of the downpour would have rushed by us, muddy with the spoils of the rocks, coming together into torrents, cutting deep gorges and canyons as they hurried past to deposit their sediment in the earliest seas. Through the clouds we should have glimpsed a great sun moving visibly across the sky, and in its wake and in the wake of the moon would have come a diurnal tide of earthquake and upheaval. And the moon, which nowadays keeps one constant face to earth, would then have been rotating visibly and showing the side it now hides so inexorably.

The earth aged. One million years followed another, and the day lengthened, the sun grew more distant and milder, the moon's pace in the sky slackened; the intensity of rain and storm diminished and the water in the first seas increased and ran together into the ocean garment our planet henceforth wore.

But there was no life as yet upon the earth; the seas were lifeless, and the rocks were barren.

III

The Beginnings of Life

As everybody knows nowadays, the knowledge we possess of life before the beginnings of human memory and tradition is derived from the markings and fossils of living things in the stratified rocks. We find preserved in shale and slate, limestone, and sandstone, bones, shells, fibres, stems, fruits, footmarks, scratchings and the like, side by side with the ripple marks of the earliest tides and the pittings of the earliest rain-falls. It is by the sedulous examination of this Record of the Rocks that the past history of the earth's life has been pieced together. That much nearly everybody knows to-day. The sedimentary rocks do not lie neatly stratum above stratum; they have been crumpled, bent, thrust about, distorted and mixed together like the leaves of a library that has been repeatedly looted and burnt, and it is only as a result of many devoted lifetimes of work that the record has been put into order and read. The whole compass of time represented by the record of the rocks is now estimated as 1,600,000,000 years.

The earliest rocks in the record are called by geologists the Azoic rocks, because they show no traces of life. Great areas of these Azoic rocks lie uncovered in North America, and they are of such a thickness that geologists consider that they represent a period of at least half of the 1,600,000,000 which they assign to the whole geological record. Let me repeat this profoundly significant fact. Half the great interval of time since land and sea were first distinguishable on earth has left us no traces of life. There are ripplings and rain marks still to be found in these rocks, but no marks nor vestiges of any living thing.

Then, as we come up the record, signs of past life appear and increase. The age of the world's history in which we find these past

MARINE LIFE IN THE CAMBRIAN PERIOD

1 and 8, Jellyfishes; 2, Hyolithes (swimming snail); 3, Hymenocaris; 4, Protospongia; 5, Lampshells (Obolella); 6, Orthoceras; 7, Trilobite (Paradoxides) — see fossil on page 13; 9, Coral (Archæocyathus); 10, Bryograptus; 11, Trilobite (Olenellus); 12, Palesterina

traces is called by geologists the Lower Palæozoic age. The first indications that life was astir are vestiges of comparatively simple and lowly things: the shells of small shellfish, the stems and flower-like heads of zoophytes, seaweeds and the tracks and remains of sea worms and crustacea. Very early appear certain creatures rather like plant-lice, crawling creatures which could roll themselves up into balls as the plant-lice do, the trilobites. Later by a few million years or so come certain sea scorpions, more mobile and powerful creatures than the world had ever seen before.

Photo: John J. Ward, F.E.S.

FOSSIL TRILOBITE (SLIGHTLY MAGNIFIED)

None of these creatures were of very great size. Among the largest were certain of the sea scorpions, which measured nine feet in length. There are no signs whatever of land life of any sort, plant or animal; there are no fishes nor any vertebrated creatures in this part of the record. Essentially all the plants and creatures which have left us their traces from this period of the earth's history are shallow-water and intertidal beings. If we wished to parallel the flora and fauna of the Lower Palæozoic rocks on the earth to-day, we should do it best, except in the matter of size, by taking a drop of water from a rock pool or scummy ditch and examining it under a microscope. The little crustacea, the small shellfish, the zoophytes and algæ we should find there would display a quite striking resemblance to these clumsier, larger prototypes that once were the crown of life upon our planet.

It is well, however, to bear in mind that the Lower Palæozoic rocks probably do not give us anything at all representative of the first beginnings of life on our planet. Unless a creature has bones or other hard parts, unless it wears a shell or is big enough and heavy enough to make characteristic footprints and trails in mud, it is unlikely to leave any fossilized traces of its existence behind. To-day there are hundreds of thousands of species of small soft-bodied creatures in our world which it is inconceivable can ever

EARLY PALÆOZOIC FOSSILS OF VARIOUS SPECIES OF LINGULA
Species of this most ancient genus of shellfish still live to-day
(*In Natural History Museum, London*)

leave any mark for future geologists to discover. In the world's past, millions of millions of species of such creatures may have lived and multiplied and flourished and passed away without a trace remaining. The waters of the warm and shallow lakes and seas of the so-called Azoic period may have teemed with an infinite variety

FOSSILIZED FOOTPRINTS OF A LABYRINTHODONT CHEIROTHERIUM
(In Natural History Museum, London)

of lowly, jelly-like, shell-less and boneless creatures, and a multitude of green scummy plants may have spread over the sunlit intertidal rocks and beaches. The Record of the Rocks is no more a complete record of life in the past than the books of a bank are a record of the existence of everybody in the neighbourhood. It is only when a species begins to secrete a shell or a spicule or a carapace or a lime-supported stem, and so put by something for the future, that it goes upon the Record. But in rocks of an age prior to those which bear any fossil traces, graphite, a form of uncombined carbon, is sometimes found, and some authorities consider that it may have been separated out from combination through the vital activities of unknown living things.

IV

The Age of Fishes

IN the days when the world was supposed to have endured for only a few thousand years, it was supposed that the different species of plants and animals were fixed and final; they had all been created exactly as they are to-day, each species by itself. But as men began to discover and study the Record of the Rocks this belief gave place to the suspicion that many species had changed and developed slowly through the course of ages, and this again expanded into a belief in what is called Organic Evolution, a belief that all species of life upon earth, animal and vegetable alike, are descended by slow continuous processes of change from some very simple ancestral form of life, some almost structureless living substance, far back in the so-called Azoic seas.

This question of Organic Evolution, like the question of the age of the earth, has in the past been the subject of much bitter controversy. There was a time when a belief in organic evolution was for rather obscure reasons supposed to be incompatible with sound Christian, Jewish and Moslem doctrine. That time has passed, and the men of the most orthodox Catholic, Protestant, Jewish and Mohammedan belief are now free to accept this newer and broader view of a common origin of all living things. No life seems to have happened suddenly upon earth. Life grew and grows. Age by age through gulfs of time at which imagination reels, life has been growing from a mere stirring in the intertidal slime towards freedom, power and consciousness.

Life consists of individuals. These individuals are definite things, they are not like the lumps and masses, nor even the limitless and motionless crystals, of non-living matter, and they have two characteristics no dead matter possesses. They can assimilate other matter into themselves and make it part of themselves, and they can reproduce themselves. They eat and they breed. They can give rise to other individuals, for the most part like themselves, but always also a little different from themselves. There is a specific and family resemblance between an individual and its offspring, and there is an individual difference between every parent and every offspring it produces, and this is true in every species and at every stage of life.

SPECIMEN OF THE PTERICHTHYS MILLERI OR SEA SCORPION SHOWING BODY ARMOUR

Now scientific men are not able to explain to us either why offspring should resemble nor why they should differ from their parents. But seeing that offspring do at once resemble and differ, it is a matter rather of common sense than of scientific knowledge that, if the conditions under which a species live are changed, the species should undergo some correlated changes. Because in any generation of the species there must be a number of individuals whose individual differences make them better adapted to the new conditions under which the species has to live, and a number whose individual differences make it rather harder for them to live. And on the whole the former sort will live longer, bear more offspring, and reproduce themselves more abundantly than the latter, and so generation by generation the average of the species will change in the favourable direction. This process, which is called Natural Selection, is not so much a scientific theory as a necessary deduction from the facts of reproduction and individual difference. There may be many forces at work varying, destroying and preserving species, about which science may still be unaware or undecided, but the man who can deny the operation of this process of natural selection upon life since its beginning must be either ignorant of the elementary facts of life or incapable of ordinary thought.

Many scientific men have speculated about the first beginning of life and their speculations are often of great interest, but there is absolutely no definite knowledge and no convincing guess yet of the way in which life began. But nearly all authorities are agreed that it probably began upon mud or sand in warm sunlit shallow brackish water, and that it spread up the beaches to the intertidal lines and out to the open waters.

Nat. Hist. Mus.

FOSSIL OF THE CLADOSELACHE, A DEVONIAN SHARK

That early world was a world of strong tides and currents. An incessant destruction of individuals must have been going on through their being swept up the beaches and dried, or by their being swept out to sea and sinking down out of reach of air and sun. Early conditions favoured the development of every tendency to root and hold on, every tendency to form an outer skin and casing to protect the stranded individual from immediate desiccation. From the very earliest any tendency to sensitiveness to taste would turn the individual in the direction of food, and any sensitiveness to light would assist it to struggle back out of the darkness of the sea deeps and caverns or to wriggle back out of the excessive glare of the dangerous shallows.

Probably the first shells and body armour of living things were protections against drying rather than against active enemies. But tooth and claw come early into our earthly history.

We have already noted the size of the earlier water scorpions. For long ages such creatures were the supreme lords of life. Then in a division of these Palæozoic rocks called the Silurian division, which many geologists now suppose to be as old as five hundred million years, there appears a new type of being, equipped with eyes and teeth and swimming powers of an altogether more powerful

By Alice Woodward

SHARKS AND GANOIDS OF THE DEVONIAN PERIOD

kind. These were the first known backboned animals, the earliest fishes, the first known Vertebrata.

These fishes increase greatly in the next division of rocks, the rocks known as the Devonian system. They are so prevalent that this period of the Record of the Rocks has been called the Age of

Fishes. Fishes of a pattern now gone from the earth, and fishes allied to the sharks and sturgeons of to-day, rushed through the waters, leapt in the air, browsed among the seaweeds, pursued and preyed upon one another, and gave a new liveliness to the waters of the world. None of these were excessively big by our present standards. Few of them were more than two or three feet long, but there were exceptional forms which were as long as twenty feet.

We know nothing from geology of the ancestors of these fishes. They do not appear to be related to any of the forms that preceded them. Zoologists have the most interesting views of their ancestry, but these they derive from the study of the development of the eggs of their still living relations, and from other sources. Apparently the ancestors of the vertebrata were soft-bodied and perhaps quite small swimming creatures who began first to develop hard parts as teeth round and about their mouths. The teeth of a skate or dogfish cover the roof and floor of its mouth and pass at the lip into the flattened toothlike scales that encase most of its body. As the fishes develop these teeth scales in the geological record, they swim out of the hidden darkness of the past into the light, the first vertebrated animals visible in the record.

V

The Age of the Coal Swamps

THE land during this Age of Fishes was apparently quite lifeless. Crags and uplands of barren rock lay under the sun and rain. There was no real soil — for as yet there were no earthworms which help to make a soil, and no plants to break up the rock particles into mould; there was no trace of moss or lichen. Life was still only in the sea.

Over this world of barren rock played great changes of climate. The causes of these changes of climate were very complex and they have still to be properly estimated. The changing shape of the earth's orbit, the gradual shifting of the poles of rotation, changes in the shapes of the continents, probably even fluctuations in the warmth of the sun, now conspired to plunge great areas of the earth's surface into long periods of cold and ice and now again for millions of years spread a warm or equable climate over this planet. There seem to have been phases of great internal activity in the world's history, when in the course of a few million years accumulated upthrusts would break out in lines of volcanic eruption and upheaval and rearrange the mountain and continental outlines of the globe, increasing the depth of the sea and the height of the mountains and exaggerating the extremes of climate. And these would be followed by vast ages of comparative quiescence, when frost, rain and river would wear down the mountain heights and carry great masses of silt to fill and raise the sea bottoms and spread the seas, ever shallower and wider, over more and more of the land. There have been "high and deep" ages in the world's history and "low and level" ages. The reader must dismiss from his mind any idea that the surface of the earth has been growing steadily cooler since its crust grew solid. After that much cooling had been achieved, the internal temperature ceased to affect surface conditions. There are traces of periods of superabundant ice and snow, of "Glacial Ages," that is, even in the Azoic period.

It was only towards the close of the Age of Fishes, in a period of extensive shallow seas and lagoons, that life spread itself out in any effectual way from the waters on to the land. No doubt

A CARBONIFEROUS SWAMP
A Coal Seam in the Making

the earlier types of the forms that now begin to appear in great abundance had already been developing in a rare and obscure manner for many scores of millions of years. But now came their opportunity.

Plants no doubt preceded animal forms in this invasion of the land, but the animals probably followed up the plant emigration very closely. The first problem that the plant had to solve was the problem of some sustaining stiff support to hold up its fronds to the sunlight when the buoyant water was withdrawn; the second was the problem of getting water from the swampy ground below to the tissues of the plant, now that it was no longer close at hand. The two problems were solved by the development of woody tissue which both sustained the plant and acted as water carrier to the leaves. The Record of the Rocks is suddenly crowded by a vast variety of woody swamp plants, many of them of great size, big tree mosses, tree ferns, gigantic horsetails and the like. And with these, age by age, there crawled out of the water a great variety of animal forms. There were centipedes and millipedes; there were the first primitive insects; there were creatures related to the ancient king crabs and sea scorpions which became the earliest spiders and land scorpions, and presently there were vertebrated animals.

Nat. Hist. Mus.

SKULL OF A LABYRINTHODONT, CAPITOSAURUS

Some of the earlier insects were very large. There were dragon flies in this period with wings that spread out to twenty-nine inches.

In various ways these new orders and genera had adapted themselves to breathing air. Hitherto all animals had breathed air dissolved in water, and that indeed is what all animals still have to do. But now in divers fashions the animal kingdom was acquiring the power of supplying its own moisture where it was needed. A man with a perfectly dry lung would suffocate to-day;

his lung surfaces must be moist in order that air may pass through them into his blood. The adaptation to air breathing consists in all cases either in the development of a cover to the old-fashioned gills to stop evaporation, or in the development of tubes or other new breathing organs lying deep inside the body and moistened by a watery secretion. The old gills with which the ancestral fish of the vertebrated line had breathed were inadaptable to breathing upon land, and in the case of this division of the animal kingdom it is the swimming bladder of the fish which becomes a new, deep-seated breathing organ, the lung. The kind of animals known as amphibia, the frogs and newts of to-day, begin their lives in the

SKELETON OF A LABYRINTHODONT: THE ERYOPS *Nat. Hist. Mus.*

water and breathe by gills; and subsequently the lung, developing in the same way as the swimming bladder of many fishes do, as a baglike outgrowth from the throat, takes over the business of breathing, the animal comes out on land, and the gills dwindle and the gill slits disappear. (All except an outgrowth of one gill slit, which becomes the passage of the ear and ear-drum.) The animal can now live only in the air, but it must return at least to the edge of the water to lay its eggs and reproduce its kind.

All the air-breathing vertebrata of this age of swamps and plants belonged to the class amphibia. They were nearly all of them forms related to the newts of to-day, and some of them attained a considerable size. They were land animals, it is true, but they were land animals needing to live in and near moist and swampy places, and all the great trees of this period were equally amphibious in their habits. None of them had yet developed fruits and seeds of a kind that could fall on land and develop with the help only of such moisture as dew and rain could bring. They all had to shed their spores in water, it would seem, if they were to germinate.

It is one of the most beautiful interests of that beautiful science, comparative anatomy, to trace the complex and wonderful adaptations of living things to the necessities of existence in air. All living things, plants and animals alike, are primarily water things. For example all the higher vertebrated animals above the fishes, up to and including man, pass through a stage in their development in the egg or before birth in which they have gill slits which are obliterated before the young emerge. The bare, water-washed eye of the fish is protected in the higher forms from drying up by eyelids and glands which secrete moisture. The weaker sound vibrations of air necessitate an ear-drum. In nearly every organ of the body similar modifications and adaptations are to be detected, similar patchings-up to meet aerial conditions.

This Carboniferous age, this age of the amphibia, was an age of life in the swamps and lagoons and on the low banks among these waters. Thus far life had now extended. The hills and high lands were still quite barren and lifeless. Life had learnt to breathe air indeed, but it still had its roots in its native water; it still had to return to the water to reproduce its kind.

VI

The Age of Reptiles

THE abundant life of the Carboniferous period was succeeded by a vast cycle of dry and bitter ages. They are represented in the Record of the Rocks by thick deposits of sandstones and the like, in which fossils are comparatively few. The temperature of the world fluctuated widely, and there were long periods of glacial cold. Over great areas the former profusion of swamp vegetation ceased, and, overlaid by these newer deposits, it began that process of compression and mineralization that gave the world most of the coal deposits of to-day.

But it is during periods of change that life undergoes its most rapid modifications, and under hardship that it learns its hardest lessons. As conditions revert towards warmth and moisture again we find a new series of animal and plant forms established. We find in the record the remains of vertebrated animals that laid eggs which, instead of hatching out tadpoles which needed to live for a time in water, carried on their development before hatching to a stage so nearly like the adult form that the young could live in air from the first moment of independent existence. Gills had been cut out altogether, and the gill slits only appeared as an embryonic phase.

These new creatures without a tadpole stage were the Reptiles. Concurrently there had been a development of seed-bearing trees, which could spread their seed, independently of swamp or lakes. There were now palmlike cycads and many tropical conifers, though as yet there were no flowering plants and no grasses. There was a great number of ferns. And there was now also an increased variety of insects. There were beetles, though bees and butterflies had yet to come. But all the fundamental forms of a new real land fauna and flora had been laid down during these vast ages of severity.

This new land life needed only the opportunity of favourable conditions to flourish and prevail.

Age by age and with abundant fluctuations that mitigation came. The still incalculable movements of the earth's crust, the changes in its orbit, the increase and diminution of the mutual inclination of orbit and pole, worked together to produce a great spell of widely diffused warm conditions. The period lasted altogether, it is now supposed, upwards of two hundred million years. It is called the

Nat. Hist. Mus.

A FOSSIL ICHTHYOSAURUS, A MESOZOIC FISH-LIZARD
Found in the Lower Lias in Somersetshire

Mesozoic period, to distinguish it from the altogether vaster Palæozoic and Azoic periods (together fourteen hundred millions) that preceded it, and from the Cainozoic or new life period that intervened between its close and the present time, and it is also called the Age of Reptiles because of the astonishing predominance and variety of this form of life. It came to an end some eighty million years ago.

In the world to-day the genera of Reptiles are comparatively few and their distribution is very limited. They are more various, it is true, than are the few surviving members of the order of the amphibia which once in the Carboniferous period ruled the world. We still have the snakes, the turtles and tortoises (the Chelonia), the alligators and crocodiles, and the lizards. Without exception they are creatures requiring warmth all the year round; they cannot stand exposure to cold, and it is probable that all the reptilian beings of the Mesozoic suffered under the same limitation. It was a hothouse fauna, living amidst a hothouse flora. It endured no frosts. But the world had at least attained a real dry land fauna and flora as distinguished from the mud and swamp fauna and flora of the previous heyday of life upon earth.

All the sorts of reptile we know now were much more abundantly represented then, great turtles and tortoises, big crocodiles and many lizards and snakes, but in addition there was a number of series of

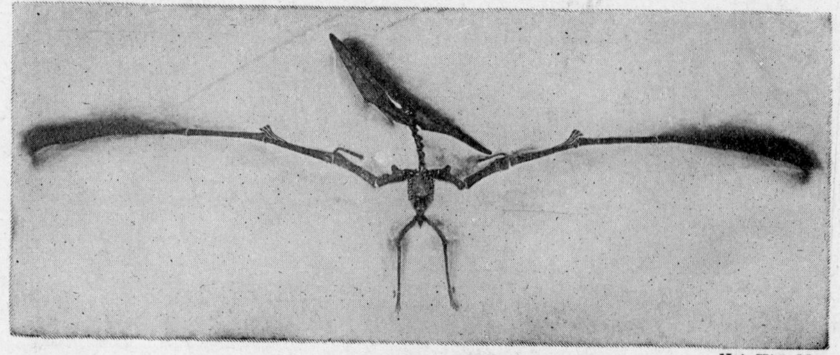

A PTERODACTYL *Nat. Hist. Mus.*

wonderful creatures that have now vanished altogether from the earth. There was a vast variety of beings called the Dinosaurs. Vegetation was now spreading over the lower levels of the world, reeds, brakes of fern and the like; and browsing upon this abundance came a multitude of herbivorous reptiles, which increased in size as the Mesozoic period rose to its climax. Some of these beasts exceeded in size any other land animals that have ever lived; they were as large as whales. The *Diplodocus Carnegii* for example measured eighty-four feet from snout to tail; the Gigantosaurus was even greater; it measured a hundred feet. Living upon these monsters was a swarm of carnivorous Dinosaurs of a corresponding size. One of these, the Tyrannosaurus, is figured and described in many books as the last word in reptilian frightfulness.

Nat. Hist. Mus.

A BIG SWAMP-INHABITING DINOSAUR, THE DIPLODOCUS, OVER EIGHTY FEET FROM SNOUT TO TAIL-TIP

While these great creatures pastured and pursued amidst the fronds and evergreens of the Mesozoic jungles, another now vanished tribe of reptiles, with a bat-like development of the fore limbs, pursued insects and one another, first leapt and parachuted and presently flew amidst the fronds and branches of the forest trees. These were the Pterodactyls. These were the first flying creatures with backbones; they mark a new achievement in the growing powers of vertebrated life.

Moreover some of the reptiles were returning to the sea waters. Three groups of big swimming beings had invaded the sea from which their ancestors had come: the Mososaurs, the Plesiosaurs, and Ichthyosaurs. Some of these again approached the proportions of our present whales. The Ichthyosaurs seem to have been quite seagoing creatures, but the Plesiosaurs were a type of animal that has no cognate form to-day. The body was stout and big with paddles, adapted either for swimming or crawling through marshes, or along the bottom of shallow waters. The comparatively small head was poised on a vast snake of neck, altogether outdoing the neck of the swan. Either the Plesiosaur swam and searched for food under the water and fed as the swan will do, or it lurked under water and snatched at passing fish or beast.

Such was the predominant land life throughout the Mesozoic age. It was by our human standards an advance upon anything that had preceded it. It had produced land animals greater in size, range, power and activity, more "vital" as people say, than anything the world had seen before. In the seas there had been no such advance but a great proliferation of new forms of life. An enormous variety of squid-like creatures with chambered shells, for the most part coiled, had appeared in the shallow seas, the Ammonites. They had had predecessors in the Palæozoic seas, but now was their age of glory. To-day they have left no survivors at all; their nearest relation is the pearly Nautilus, an inhabitant of tropical waters. And a new and more prolific type of fish with lighter, finer scales than the plate-like and tooth-like coverings that had hitherto prevailed, became and has since remained predominant in the seas and rivers.

VII

The First Birds and the First Mammals

IN a few paragraphs a picture of the lush vegetation and swarming reptiles of that first great summer of life, the Mesozoic period, has been sketched. But while the Dinosaurs lorded it over the hot selvas and marshy plains and the Pterodactyls filled the forests with their flutterings and possibly with shrieks and croakings as they pursued the humming insect life of the still flowerless shrubs and trees, some less conspicuous and less abundant forms upon the margins of this abounding life were acquiring certain powers and learning certain lessons of endurance, that were to be of the utmost value to their race when at last the smiling generosity of sun and earth began to fade.

A group of tribes and genera of hopping reptiles, small creatures of the dinosaur type, seem to have been pushed by competition and the pursuit of their enemies towards the alternatives of extinction or adaptation to colder conditions in the higher hills or by the sea. Among these distressed tribes there was developed a new type of scale — scales that were elongated into quill-like forms and that presently branched into the crude beginnings of feathers. These quill-like scales lay over one another and formed a heat-retaining covering more efficient than any reptilian covering that had hitherto existed. So they permitted an invasion of colder regions that were otherwise uninhabited. Perhaps simultaneously with these changes there arose in these creatures a greater solicitude for their eggs. Most reptiles are apparently quite careless about their eggs, which are left for sun and season to hatch. But some of the varieties upon this new branch of the tree of life were acquiring a habit of guarding their eggs and keeping them warm with the warmth of their bodies.

With these adaptations to cold other internal modifications

Nat. Hist. Mus.

FOSSIL OF THE ARCHEOPTERYX; ONE OF THE EARLIEST BIRDS

were going on that made these creatures, the primitive birds, warm-blooded and independent of basking. The very earliest birds seem to have been sea-birds living upon fish, and their fore limbs were not wings but paddles rather after the penguin type. That peculiarly primitive bird, the New Zealand Ki-wi, has feathers of a very simple sort, and neither flies nor appears to be descended from flying ancestors. In the development of the birds, feathers came before wings. But once the feather was developed the possibility of making a light spread of feathers led inevitably to the wing. We know of the fossil remains of one bird at least which had reptilian teeth in its jaw and a long reptilian tail, but which also had a true bird's wing and which certainly flew and held its own among the pterodactyls of the Mesozoic time. Nevertheless birds were neither varied nor abundant in Mesozoic times. If a man could go back to typical Mesozoic country, he might walk for days and never see or hear such a thing as a bird, though he would see a great abundance of pterodactyls and insects among the fronds and reeds.

And another thing he would probably never see, and that would be any sign of a mammal. Probably the first mammals were in existence millions of years before the first thing one could call a bird, but they were altogether too small and obscure and remote for attention.

HESPERORNIS IN ITS NATIVE SEAS

The earliest mammals, like the earliest birds, were creatures driven by competition and pursuit into a life of hardship and adaptation to cold. With them also the scale became quill-like, and was developed into a heat-retaining covering; and they too underwent modifications, similar in kind though different in detail, to become warm-blooded and independent of basking. Instead of feathers they developed hairs, and instead of guarding and incubating their eggs they kept them warm and safe by retaining them inside their bodies until they were almost mature. Most of them became altogether vivaparous and brought their young into the world alive. And even after their young were born they tended to maintain a protective and nutritive association with them. Most but not all mammals to-day have mammæ and suckle their young. Two mammals still live which lay eggs and which have not proper mammæ, though they nourish their young by a nutritive secretion of the under skin; these are the duck-billed platypus and the echidna. The echidna lays leathery eggs and then puts them into a pouch under its belly, and so carries them about warm and safe until they hatch.

But just as a visitor to the Mesozoic world might have searched for days and weeks before finding a bird, so, unless he knew exactly where to go and look, he might have searched in vain for any traces of a mammal. Both birds and mammals would have seemed very eccentric and secondary and unimportant creatures in Mesozoic times.

The Age of Reptiles lasted, it is now guessed, eighty million years. Had any quasi-human intelligence been watching the world through that inconceivable length of time, how safe and eternal the sunshine and abundance must have seemed, how assured the wallowing prosperity of the dinosaurs and the flapping abundance of the flying lizards! And then the mysterious rhythms and accumulating forces of the universe began to turn against that quasi-eternal stability. That run of luck

Photo: Autotype Fine Art Co.

THE KI-WI, APTERYX, STILL FOUND IN NEW ZEALAND

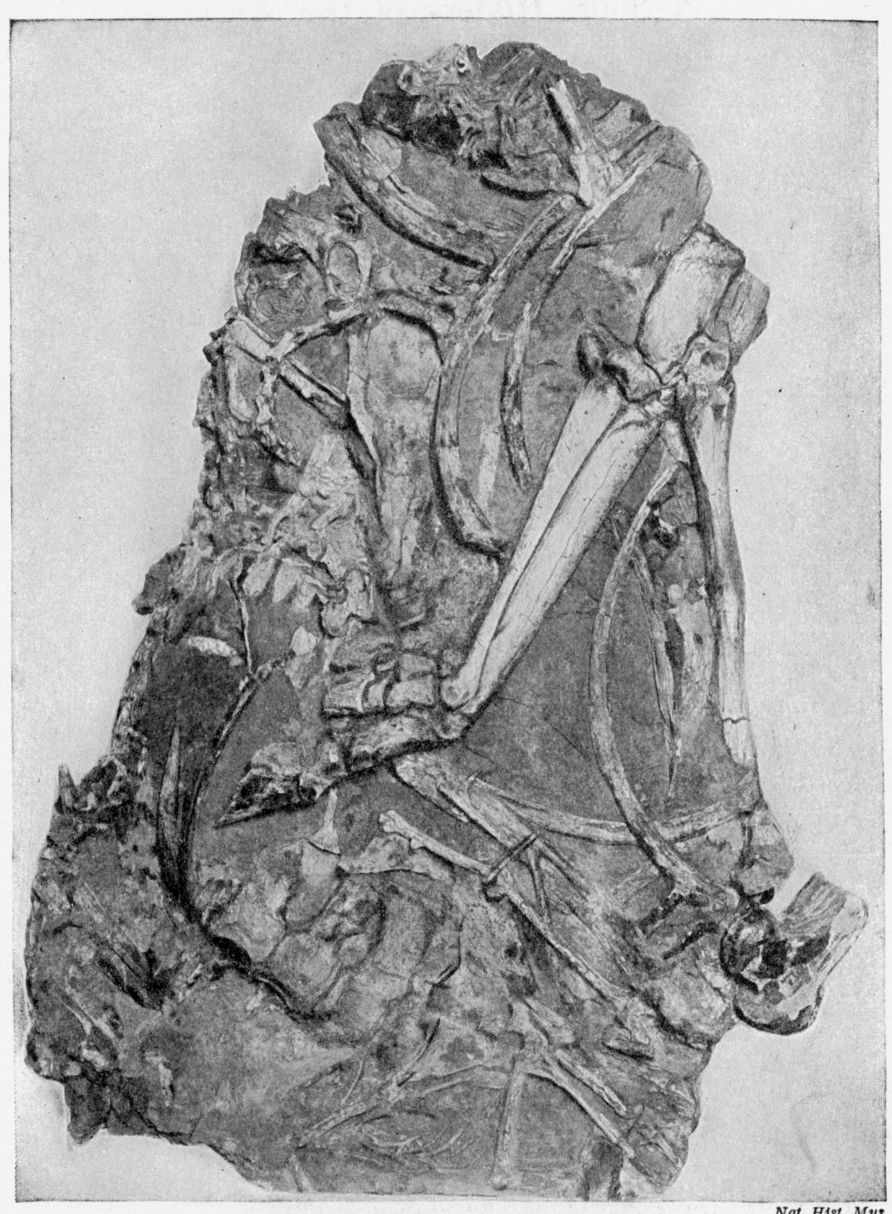

Nat. Hist. Mus.

SLAB OF LOWER PLIOCENE MARL

Discovered in Greece; it is rich in fossilized bones of early mammals

for life was running out. Age by age, myriad of years after myriad of years, with halts no doubt and retrogressions, came a change towards hardship and extreme conditions, came great alterations of level and great redistributions of mountain and sea. We find one thing in the Record of the Rocks during the decadence of the long Mesozoic age of prosperity that is very significant of steadily sustained changes of condition, and that is a violent fluctuation of living forms and the appearance of new and strange species. Under the gathering threat of extinction the older orders and genera are displaying their utmost capacity for variation and adaptation. The Ammonites for example in these last pages of the Mesozoic chapter exhibit a multitude of fantastic forms. Under settled conditions there is no encouragement for novelties; they do not develop, they are suppressed; what is best adapted is already there. Under novel conditions it is the ordinary type that suffers, and the novelty that may have a better chance to survive and establish itself. . . .

There comes a break in the Record of the Rocks that may represent several million years. There is a veil here still, over even the outline of the history of life. When it lifts again, the Age of Reptiles is at an end; the Dinosaurs, the Plesiosaurs and Ichthyosaurs, the Pterodactyls, the innumerable genera and species of Ammonite have all gone absolutely. In all their stupendous variety they have died out and left no descendants. The cold has killed them. All their final variations were insufficient; they had never hit upon survival conditions. The world had passed through a phase of extreme conditions beyond their powers of endurance, a slow and complete massacre of Mesozoic life has occurred, and we find now a new scene, a new and hardier flora, and a new and hardier fauna in possession of the world.

It is still a bleak and impoverished scene with which this new volume of the book of life begins. The cycads and tropical conifers have given place very largely to trees that shed their leaves to avoid destruction by the snows of winter and to flowering plants and shrubs, and where there was formerly a profusion of reptiles, an increasing variety of birds and mammals is entering into their inheritance.

VIII

The Age of Mammals

THE opening of the next great period in the life of the earth, the Cainozoic period, was a period of upheaval and extreme volcanic activity. Now it was that the vast masses of the Alps and Himalayas and the mountain backbone of the Rockies and Andes were thrust up, and that the rude outlines of our present oceans and continents appeared. The map of the world begins to display a first dim resemblance to the map of to-day. It is estimated now that between forty and eighty million years have elapsed from the beginnings of the Cainozoic period to the present time.

At the outset of the Cainozoic period the climate of the world was austere. It grew generally warmer until a fresh phase of great abundance was reached, after which conditions grew hard again and the earth passed into a series of extremely cold cycles, the Glacial Ages, from which apparently it is now slowly emerging.

But we do not know sufficient of the causes of climatic change at present to forecast the possible fluctuations of climatic conditions that lie before us. We may be moving towards increasing sunshine or lapsing towards another glacial age; volcanic activity and the upheaval of mountain masses may be increasing or diminishing; we do not know; we lack sufficient science.

With the opening of this period the grasses appear; for the first time there is pasture in the world; and with the full development of the once obscure mammalian type, appear a number of interesting grazing animals and of carnivorous types which prey upon these.

At first these early mammals seem to differ only in a few characters from the great herbivorous and carnivorous reptiles that ages before had flourished and then vanished from the earth. A careless observer might suppose that in this second long age of warmth and plenty that was now beginning, nature was merely repeating the first, with herbivorous and carnivorous mammals to parallel the herbivorous and carnivorous dinosaurs, with birds replacing pterodactyls and so on. But this would be an altogether superficial comparison. The variety of the universe is infinite and incessant; it progresses eternally; history never repeats itself and no parallels are precisely true. The differences between the life of the Cainozoic and Mesozoic periods are far profounder than the resemblances.

The most fundamental of all these differences lies in the mental life of the two periods. It arises essentially out of the continuing contact of parent and offspring which distinguishes mammalian and in a lesser degree bird life, from the life of the reptile. With very few exceptions the reptile abandons its egg to hatch alone. The young reptile has no knowledge whatever of its parent; its mental life, such as it is, begins and ends with its own experiences.

A MAMMAL OF THE EARLY CAINOZOIC PERIOD
The Titanotherium (Brontops) Robustum

It may tolerate the existence of its fellows but it has no communication with them; it never imitates, never learns from them, is incapable of concerted action with them. Its life is that of an isolated individual. But with the suckling and cherishing of young which was distinctive of the new mammalian and avian strains arose the possibility of learning by imitation, of communication, by warning cries and other concerted action, of mutual control and instruction. A *teachable* type of life had come into the world.

The earliest mammals of the Cainozoic period are but little superior in brain size to the more active carnivorous dinosaurs, but as we read on through the record towards modern times we find, in every tribe and race of the mammalian animals, a steady universal increase in brain capacity. For instance we find at a comparatively early stage that rhinoceros-like beasts appear. There is a creature, the Titanotherium, which lived in the earliest division of this period. It was probably very like a modern rhinoceros in its habits and needs. But its brain capacity was not one tenth that of its living successor.

The earlier mammals probably parted from their offspring as soon as suckling was over, but, once the capacity for mutual understanding has arisen, the advantages of continuing the association are very great; and we presently find a number of mammalian species displaying the beginnings of a true social life and keeping together in herds, packs and flocks, watching each other, imitating each other, taking warning from each other's acts and cries. This is something that the world had not seen before among vertebrated animals. Reptiles and fish may no doubt be found in swarms and shoals; they have been hatched in quantities and similar conditions have kept them together, but in the case of the social and gregarious mammals the association arises not simply from a community of external forces, it is sustained by an inner impulse. They are not merely like one another and so found in the same places at the same times; they like one another and so they keep together.

This difference between the reptile world and the world of our human minds is one our sympathies seem unable to pass. We cannot conceive in ourselves the swift uncomplicated urgency of a reptile's instinctive motives, its appetites, fears and hates. We

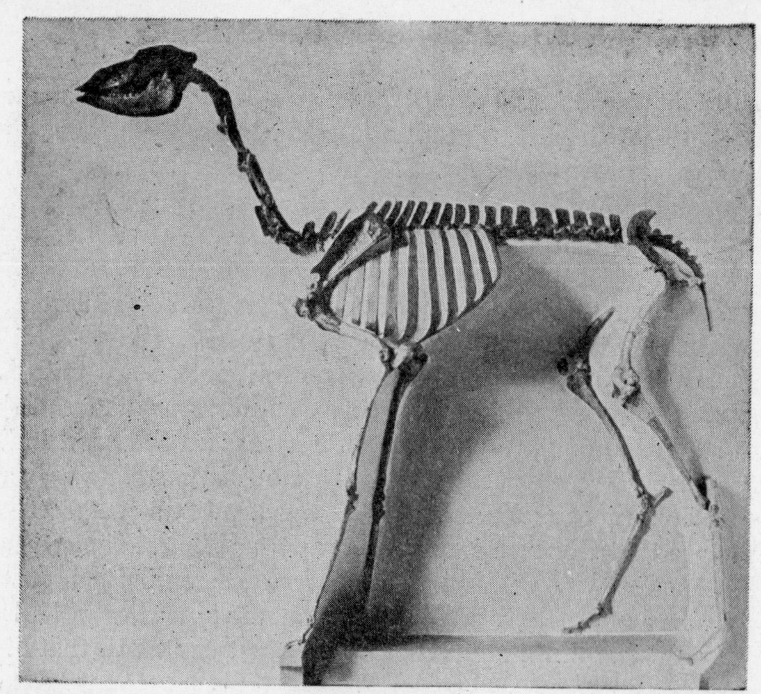

Nat. Hist. Mus.

STENOMYLUS HITCHCOCKI—A GIRAFFE-CAMEL

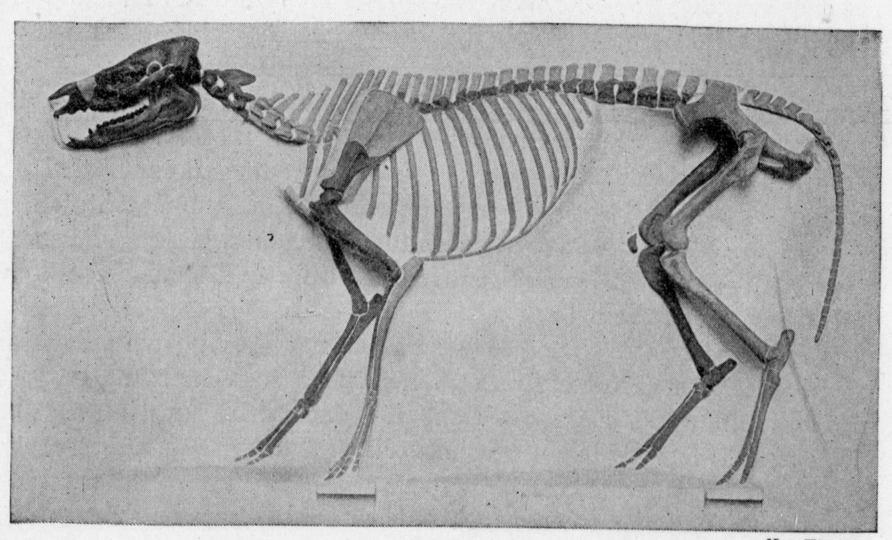

SKELETON OF PROTOHIPPUS VENTICOLUS—EARLY HORSE Nat. Hist. Mus.

cannot understand them in their simplicity because all our motives are complicated; ours are balances and resultants and not simple urgencies. But the mammals and birds have self-restraint and consideration for other individuals, a social appeal, a self-control that is, at its lower level, after our own fashion. We can in consequence establish relations with almost all sorts of them. When they suffer they utter cries and make movements that rouse our

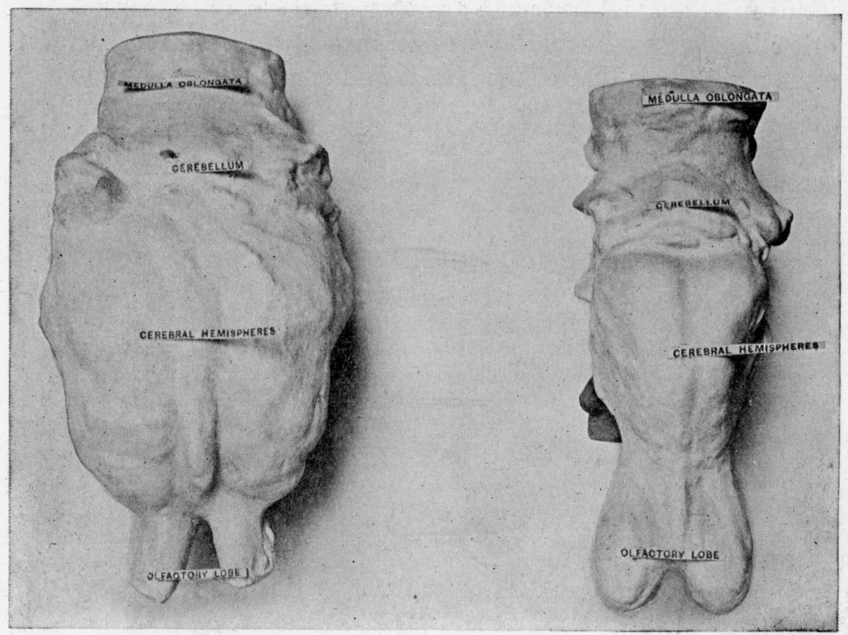

Nat. Hist. Mus.

COMPARATIVE SIZES OF BRAINS OF RHINOCEROS AND DINOCERAS

feelings. We can make understanding pets of them with a mutual recognition. They can be tamed to self-restraint towards us, domesticated and taught.

That unusual growth of brain which is the central fact of Cainozoic times marks a new communication and interdependence of individuals. It foreshadows the development of human societies of which we shall soon be telling.

As the Cainozoic period unrolled, the resemblance of its flora and fauna to the plants and animals that inhabit the world to-day increased. The big clumsy Uintatheres and Titanotheres, the Entelodonts and Hyracodons, big clumsy brutes like nothing living, disappeared. On the other hand a series of forms led up by steady degrees from grotesque and clumsy predecessors to the giraffes, camels, horses, elephants, deer, dogs and lions and tigers of the existing world. The evolution of the horse is particularly legible upon the geological record. We have a fairly complete series of forms from a small tapir-like ancestor in the early Cainozoic. Another line of development that has now been pieced together with some precision is that of the llamas and camels.

IX

MONKEYS, APES AND SUB-MEN

NATURALISTS divide the class *Mammalia* into a number of orders. At the head of these is the order *Primates*, which includes the lemurs, the monkeys, apes and man. Their classification was based originally upon anatomical resemblances and took no account of any mental qualities.

Now the past history of the Primates is one very difficult to decipher in the geological record. They are for the most part animals which live in forests like the lemurs and monkeys or in bare rocky places like the baboons. They are rarely drowned and covered up by sediment, nor are most of them very numerous species, and so they do not figure so largely among the fossils as the ancestors of the horses, camels and so forth do. But we know that quite early in the Cainozoic period, that is to say some forty million years ago or so, primitive monkeys and lemuroid creatures had appeared, poorer in brain and not so specialized as their later successors.

The great world summer of the middle Cainozoic period drew at last to an end. It was to follow those other two great summers in the history of life, the summer of the Coal Swamps and the vast summer of the Age of Reptiles. Once more the earth spun towards an ice age. The world chilled, grew milder for a time and chilled again. In the warm past hippopotami had wallowed through a lush sub-tropical vegetation, and a tremendous tiger with fangs like sabres, the sabre-toothed tiger, had hunted its prey where now the journalists of Fleet Street go to and fro. Now came a bleaker age and still bleaker ages. A great weeding and extinction of species occurred. A woolly rhinoceros, adapted to a cold climate, and the mammoth, a big woolly cousin of the elephants, the Arctic musk ox and the reindeer passed across the scene. Then century by century the Arctic ice cap, the wintry death of the great Ice Age, crept southward. In England it came almost down to the Thames, in America it reached Ohio. There would be warmer spells of a few thousand years and relapses towards a bitterer cold.

Geologists talk of these wintry phases as the First, Second, Third and Fourth Glacial Ages, and of the interludes as Interglacial periods. We live to-day in a world that is still impoverished and scarred by that terrible winter. The First Glacial Age was coming on 600,000 years ago; the Fourth Glacial Age reached its bitterest

A MAMMOTH

some fifty thousand years ago. And it was amidst the snows of this long universal winter that the first man-like beings lived upon our planet.

By the middle Cainozoic period there have appeared various apes with many quasi-human attributes of the jaws and leg bones, but it is only as we approach these Glacial Ages that we find traces of creatures that we can speak of as "almost human." These traces are not bones but implements. In Europe, in deposits of this period, between half a million and a million years old, we find flints and stones that have evidently been chipped intentionally by some handy creature desirous of hammering, scraping or fighting with the sharpened edge. These things have been called "Eoliths" (dawn stones). In Europe there are no bones nor other remains of the creature which made these objects, simply the objects themselves. For all the certainty we have it may have been some entirely unhuman but intelligent monkey. But at Trinil in Java, in accumulations of this age, a piece of a skull and various teeth and bones have been found of a sort of ape man, with a brain case bigger than that of any living apes, which seems to have walked erect. This creature is now called *Pithecanthropus erectus*, the walking ape man, and the little trayful of its bones is the only help our imaginations have as yet in figuring to ourselves the makers of the Eoliths.

Nat. Hist. Mus.

FLINT IMPLEMENTS FOUND IN PILTDOWN REGION

It is not until we come to sands that are almost a quarter of a million years old that we find any other particle of a sub-human being. But there are plenty of implements, and they are steadily improving in quality as we read on through the record. They are no longer clumsy Eoliths; they are now shapely instruments made with considerable skill. *And they are much bigger than the similar implements afterwards made by true man.* Then, in a sandpit at Heidelberg, appears a single quasi-human jaw-bone, a clumsy jaw-bone, absolutely chinless, far heavier than a true human jaw-bone and narrower, so that it is improbable the creature's tongue could have moved about for articulate speech. On the strength of this jaw-bone, scientific men suppose this creature to have been a heavy, almost human monster, possibly with huge limbs and hands, possibly with a thick felt of hair, and they call it the Heidelberg Man.

A THEORETICAL RESTORATION OF THE PITHECANTHROPUS ERECTUS BY PROF. RUTOT

THE HEIDELBERG MAN

The Heidelberg Man, as modelled under the supervision of Prof. Rutot

This jaw-bone is, I think, one of the most tormenting objects in the world to our human curiosity. To see it is like looking through a defective glass into the past and catching just one blurred and tantalizing glimpse of this Thing, shambling through the bleak wilderness, clambering to avoid the sabre-toothed tiger, watching the woolly rhinoceros in the woods. Then before we can scrutinize the monster, he vanishes. Yet the soil is littered abundantly with the indestructible implements he chipped out for his uses.

Still more fascinatingly enigmatical are the remains of a creature found at Piltdown in Sussex in a deposit that may indicate an age between a hundred and a hundred and fifty thousand years ago, though some authorities would put these particular remains back in time to before the Heidelberg jaw-bone. Here there are the remains of a thick sub-human skull much larger than any existing ape's, and a chimpanzee-like jaw-bone which may or may not belong to it, and, in addition, a bat-shaped piece of elephant bone evidently carefully manufactured, through which a hole had apparently been bored. There is also the thigh-bone of a deer with cuts upon it like a tally. That is all.

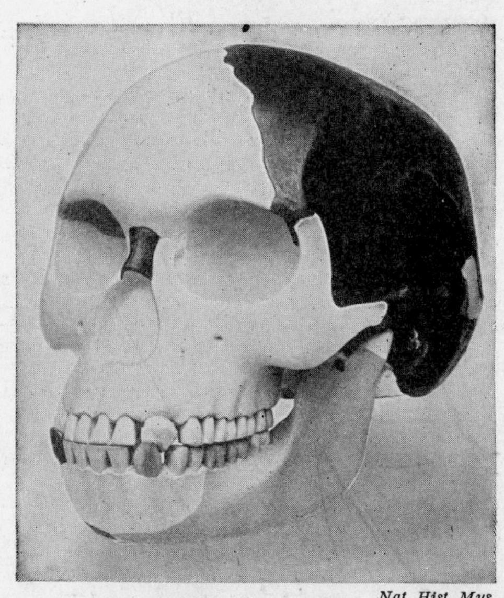

Nat. Hist. Mus.

THE PILTDOWN SKULL, AS RECONSTRUCTED FROM ORIGINAL FRAGMENT

What sort of beast was this creature which sat and bored holes in bones?

Scientific men have named him Eoanthropus, the Dawn Man. He stands apart from his kindred; a very different being either from the Heidelberg creature or from any living ape. No other vestige like him is known. But the gravels and deposits of from one hundred thousand years onward are increasingly rich in implements of flint and similar stone. And these implements are no longer rude "Eoliths." The archæologists are presently able to distinguish scrapers, borers, knives, darts, throwing stones and hand axes. . . .

We are drawing very near to man. In our next section we shall have to describe the strangest of all these precursors of humanity, the Neanderthalers, the men who were almost, but not quite, true men.

But it may be well perhaps to state quite clearly here that no scientific man supposes either of these creatures, the Heidelberg Man or *Eoanthropus,* to be direct ancestors of the men of to-day. These are, at the closest, related forms.

X

The Neanderthaler and the Rhodesian Man

ABOUT fifty or sixty thousand years ago, before the climax of the Fourth Glacial Age, there lived a creature on earth so like a man that until a few years ago its remains were considered to be altogether human. We have skulls and bones of it and a great accumulation of the large implements it made and used. It made fires. It sheltered in caves from the cold. It probably dressed skins roughly and wore them. It was right-handed as men are.

Yet now the ethnologists tell us these creatures were not true men. They were of a different species of the same genus. They had heavy protruding jaws and great brow ridges above the eyes and very low foreheads. Their thumbs were not opposable to the fingers as men's are; their necks were so poised that they could not turn back their heads and look up to the sky. They probably slouched along, head down and forward. Their chinless jaw-bones resemble the Heidelberg jaw-bone and are markedly unlike human jaw-bones. And there were great differences from the human pattern in their teeth. Their cheek teeth were more complicated in structure than ours, more complicated and not less so; they had not the long fangs of our cheek teeth; and also these quasi-men had not the marked canines (dog teeth) of an ordinary human being. The capacity of their skulls was quite human, but the brain was bigger behind and lower in front than the human brain. Their intellectual faculties were differently arranged. They were not ancestral to the human line. Mentally and physically they were upon a different line from the human line.

Skulls and bones of this extinct species of man were found at Neanderthal among other places, and from that place these strange proto-men have been christened Neanderthal Men, or Neanderthalers. They must have endured in Europe for many hundreds or even thousands of years.

At that time the climate and geography of our world was very different from what they are at the present time. Europe for example was covered with ice reaching as far south as the Thames and into Central Germany and Russia; there was no Channel separating Britain from France; the Mediterranean and the Red Sea

THE NEANDERTHALER, ACCORDING TO PROF. RUTOT

were great valleys, with perhaps a chain of lakes in their deeper portions, and a great inland sea spread from the present Black Sea across South Russia and far into Central Asia. Spain and all of Europe not actually under ice consisted of bleak uplands under a harder climate than that of Labrador, and it was only when North Africa was reached that one would have found a temperate climate. Across the cold steppes of Southern Europe with its sparse arctic vegetation, drifted such hardy creatures as the woolly mammoth, and woolly rhinoceros, great oxen and reindeer, no doubt following the vegetation northward in spring and southward in autumn.

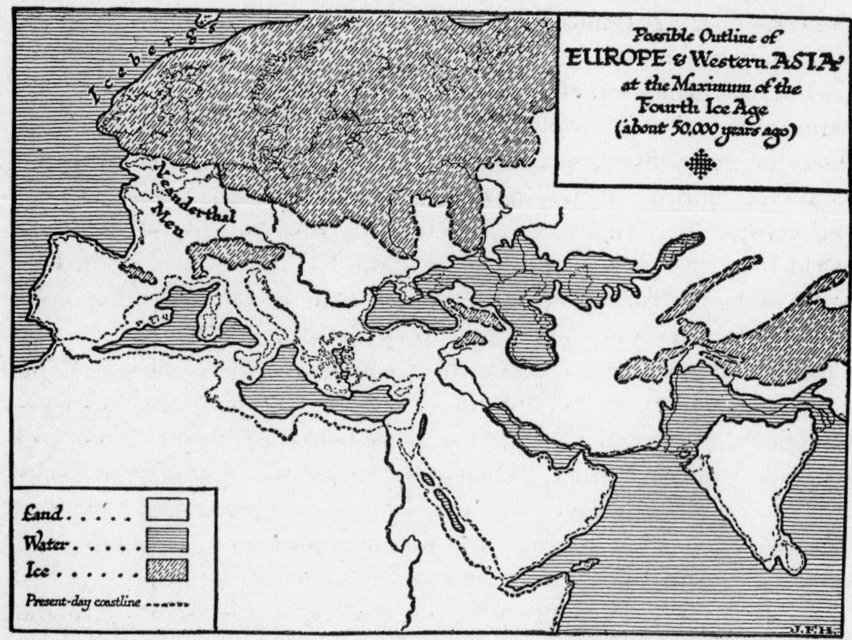

Such was the scene through which the Neanderthaler wandered, gathering such subsistence as he could from small game or fruits and berries and roots. Possibly he was mainly a vegetarian, chewing twigs and roots. His level elaborate teeth suggest a largely vegetarian dietary. But we also find the long marrow bones of great animals in his caves, cracked to extract the marrow. His weapons could not have been of much avail in open conflict with great beasts, but it is supposed that he attacked them with spears at difficult river crossings and even constructed pitfalls for them. Possibly he followed the herds and preyed upon any dead that were killed in fights, and perhaps he played the part of jackal to the sabre-toothed tiger which still survived in his day. Possibly in the bitter hardships of the Glacial Ages this creature had taken to attacking animals after long ages of vegetarian adaptation.

We cannot guess what this Neanderthal man looked like. He may have been very hairy and very inhuman-looking indeed. It is even doubtful if he went erect. He may have used his knuckles as well as his feet to hold himself up. Probably he went about alone or in small family groups. It is inferred from the structure of his jaw that he was incapable of speech as we understand it.

For thousands of years these Neanderthalers were the highest animals that the European area had ever seen; and then some thirty or thirty-five thousand years ago as the climate grew warmer a race of kindred beings, more intelligent, knowing more, talking and co-operating together, came drifting into the Neanderthaler's world from the south. They ousted the Neanderthalers from their caves and squatting places; they hunted the same food; they probably made war upon their grisly predecessors and killed them off. These newcomers from the south or the east — for at present we do not know their region of origin — who at last drove the Neanderthalers out of existence altogether, were beings of our own blood and kin, the first True Men. Their brain-cases and thumbs and necks and teeth were anatomically the same as our own. In a cave at Cro-Magnon and in another at Grimaldi, a number of skeletons have been found, the earliest truly human remains that are so far known.

So it is our race comes into the Record of the Rocks, and the story of mankind begins.

The world was growing liker our own in those days though the climate was still austere. The glaciers of the Ice Age were receding in Europe; the reindeer of France and Spain presently gave way to great herds of horses as grass increased upon the steppes, and the

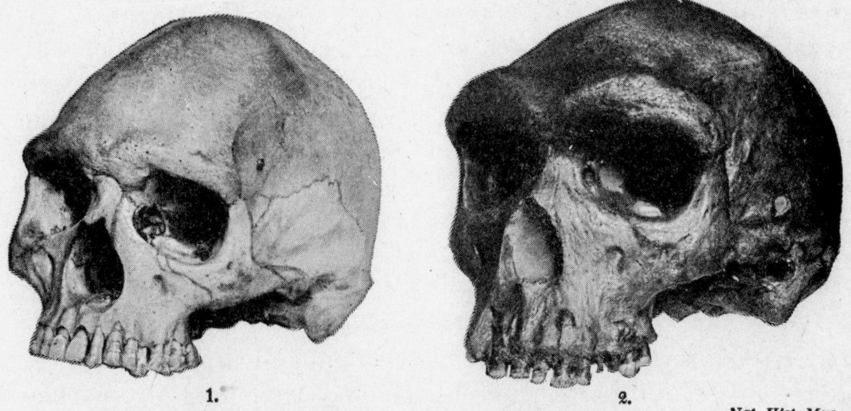

Nat. Hist. Mus.

COMPARISON OF (1) MODERN SKULL AND (2) RHODESIAN SKULL

mammoth became more and more rare in southern Europe and finally receded northward altogether. . . .

We do not know where the True Men first originated. But in the summer of 1921, an extremely interesting skull was found together with pieces of a skeleton at Broken Hill in South Africa, which seems to be a relic of a third sort of man, intermediate in its characteristics between the Neanderthaler and the human being. The brain-case indicates a brain bigger in front and smaller behind than the Neanderthaler's, and the skull was poised erect upon the backbone in a quite human way. The teeth also and the bones are quite human. But the face must have been ape-like with enormous brow ridges and a ridge along the middle of the skull. The creature was indeed a true man, so to speak, with an ape-like, Neanderthaler face. This Rhodesian Man is evidently still closer to real men than the Neanderthal Man.

This Rhodesian skull is probably only the second of what in the end may prove to be a long list of finds of sub-human species which lived on the earth in the vast interval of time between the beginnings of the Ice Age and the appearance of their common heir, and perhaps their common exterminator, the True Man. The Rhodesian skull itself may not be very ancient. Up to the time of publishing this book there has been no exact determination of its probable age. It may be that this sub-human creature survived in South Africa until quite recent times.

XI

The First True Men

THE earliest signs and traces at present known to science, of a humanity which is indisputably kindred with ourselves, have been found in western Europe and particularly in France and Spain. Bones, weapons, scratchings upon bone and rock, carved fragments of bone, and paintings in caves and upon rock surfaces dating, it is supposed, from 30,000 years ago or more, have been discovered in both these countries. Spain is at present the richest country in the world in these first relics of our real human ancestors.

Of course our present collections of these things are the merest beginnings of the accumulations we may hope for in the future, when there are searchers enough to make a thorough examination of all possible sources and when other countries in the world, now inaccessible to archæologists, have been explored in some detail. The greater part of Africa and Asia has never even been traversed yet by a trained observer interested in these matters and free to explore, and we must be very careful therefore not to conclude that the early true men were distinctively inhabitants of western Europe or that they first appeared in that region.

In Asia or Africa or submerged beneath the sea of to-day there may be richer and much earlier deposits of real human remains than anything that has yet come to light. I write in Asia or Africa, and I do not mention America because so far there have been no finds at all of any of the higher Primates, either of great apes, sub-men, Neanderthalers nor early true men. This development of life seems to have been an exclusively old world development, and it was only apparently at the end of the Old Stone Age that human beings first made their way across the land connexion that is now cut by Behring Straits, into the American continent.

These first real human beings we know of in Europe appear already to have belonged to one or other of at least two very distinct races. One of these races was of a very high type indeed; it was tall and big brained. One of the women's skulls found exceeds in capacity that of the average man of to-day. One of the men's skeletons is over six feet in height. The physical type resembled that of the North American Indian. From the Cro-Magnon cave in

ONE OF THE MARVELLOUS CAVE PAINTINGS OF ALTAMIRA, NORTH SPAIN

The Walls of the Caves are covered in these representations of Bulls, etc., painted in soft tones of red shaded to black. They may be fifteen or twenty thousand years old

which the first skeletons were found these people have been called Cro-Magnards. They were savages, but savages of a high order. The second race, the race of the Grimaldi cave remains, was distinctly negroid in its characters. Its nearest living affinities are the Bushmen and Hottentots of South Africa. It is interesting to find at the very outset of the known human story, that mankind was already racially divided into at least two main varieties; and one is tempted to such unwarrantable guesses as that the former race was probably brownish rather than black and that it came from the East or North, and that the latter was blackish rather than brown and came from the equatorial south.

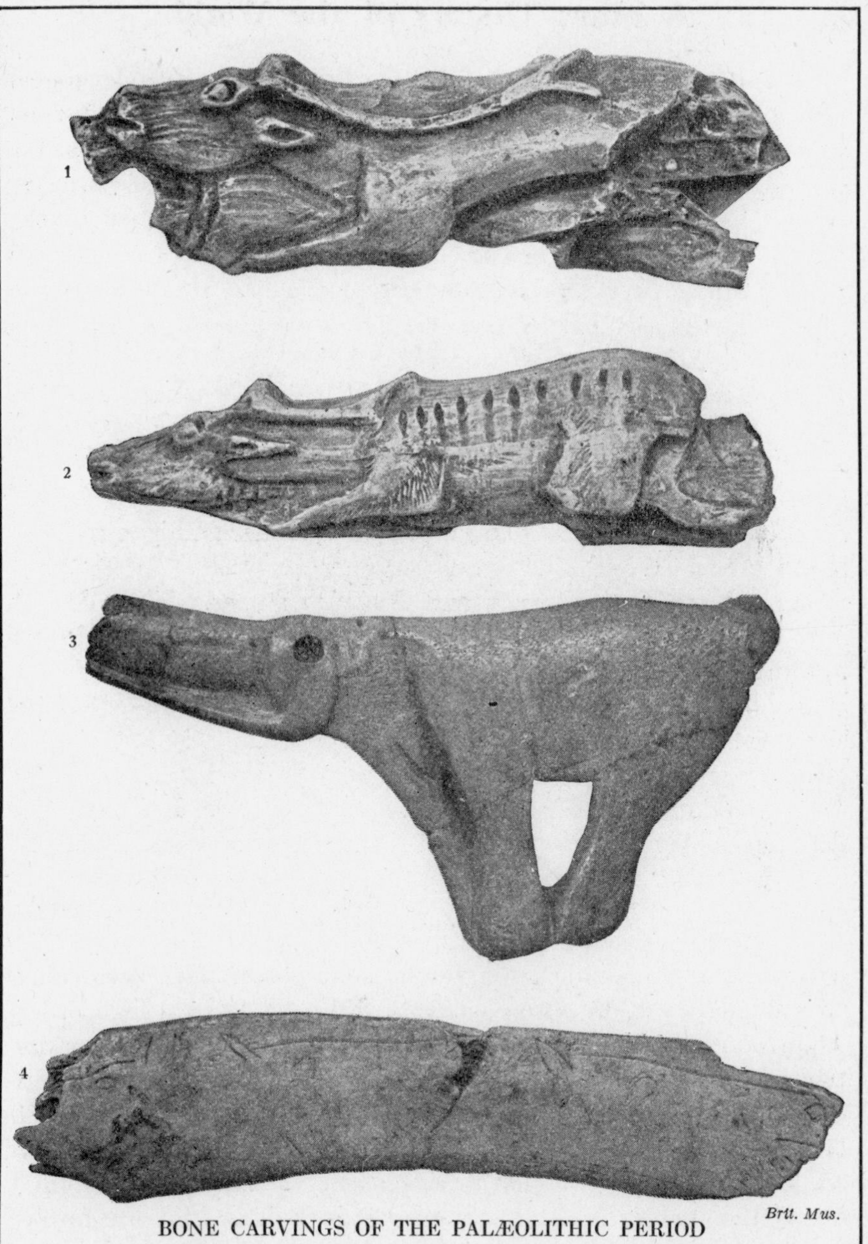

BONE CARVINGS OF THE PALÆOLITHIC PERIOD

(1 and 2) Mammoth tusk carved to shape of Reindeer, (3) Dagger Handle representing Mammoth, and (4) Bone engraved with Horses' Heads

And these savages of perhaps forty thousand years ago were so human that they pierced shells to make necklaces, painted themselves, carved images of bone and stone, scratched figures on rocks and bones, and painted rude but often very able sketches of beasts and the like upon the smooth walls of caves and upon inviting rock surfaces. They made a great variety of implements, much smaller in scale and finer than those of the Neanderthal men. We have now in our museums great quantities of their implements, their statuettes, their rock drawings and the like.

The earliest of them were hunters. Their chief pursuit was the wild horse, the little bearded pony of that time. They followed it as it moved after pasture. And also they followed the bison. They knew the mammoth, because they have left us strikingly effective pictures of that creature. To judge by one rather ambiguous drawing they trapped and killed it.

They hunted with spears and throwing stones. They do not seem to have had the bow, and it is doubtful if they had yet learnt to tame any animals. They had no dogs. There is one carving of a horse's head and one or two drawings that suggest a bridled horse, with a twisted skin or tendon round it. But the little horses of that age and region could not have carried a man, and if the horse was domesticated it was used as a led horse. It is doubtful and improbable that they had yet learnt the rather unnatural use of animal's milk as food.

They do not seem to have erected any buildings though they may have had tents of skins, and though they made clay figures they never rose to the making of pottery. Since they had no cooking implements their cookery must have been rudimentary or nonexistent. They knew nothing of cultivation and nothing of any sort of basket work or woven cloth. Except for their robes of skin or fur they were naked painted savages.

These earliest known men hunted the open steppes of Europe for a hundred centuries perhaps, and then slowly drifted and changed before a change of climate. Europe, century by century, was growing milder and damper. Reindeer receded northward and eastward, and bison and horse followed. The steppes gave way to forests, and red deer took the place of horse and bison. There is a change in the character of the implements with this change in their application. River and lake fishing becomes of great importance to men, and fine implements of bone increased. "The bone needles of this age," says de Mortillet, "are much superior to those of later, even historical times, down to the Renaissance. The Romans, for example, never had needles comparable to those of this epoch."

THE RUTOT BUST OF A CRO-MAGNON MAN

Almost fifteen or twelve thousand years ago a fresh people drifted into the south of Spain, and left very remarkable drawings of themselves upon exposed rock faces there. These were the Azilians (named from the Mas d'Azil cave). They had the bow; they seem to have worn feather headdresses; they drew vividly; but also they had reduced their drawings to a sort of symbolism — a man for instance would be represented by a vertical dab with two or three horizontal dabs — that suggest the dawn of the writing idea. Against hunting sketches there are often marks like tallies. One drawing shows two men smoking out a bees' nest.

THE HONEY GATHERER AMONG THE BEES

He is on a rope-ladder

FIGHT OF BOWMEN

Among the most recent discoveries of Palæolithic Art are these specimens found in 1920 in Spain. They are probably ten or twelve thousand years old

These are the latest of the men that we call Palæolithic (Old Stone Age) because they had only chipped implements. By ten or twelve thousand years a new sort of life has dawned in Europe, men have learnt not only to chip but to polish and grind stone implements, and they have begun cultivation. The Neolithic Age (New Stone Age) was beginning.

It is interesting to note that less than a century ago there still survived in a remote part of the world, in Tasmania, a race of human beings at a lower level of physical and intellectual development than any of these earliest races of mankind who have left traces in Europe. These people had long ago been cut off by geographical changes from the rest of the species, and from stimulation and improvement. They seem tc have degenerated rather than developed. They lived a base life subsisting upon shellfish and small game. They had no habitations but only squatting places. They were real men of our species, but they had neither the manual dexterity nor the artistic powers of the first true men.

XII

Primitive Thought

AND now let us indulge in a very interesting speculation; how did it feel to be a man in those early days of the human adventure? How did men think and what did they think in those remote days of hunting and wandering four hundred centuries ago before seed time and harvest began. Those were days long before the written record of any human impressions, and we are left almost entirely to inference and guesswork in our answers to these questions.

The sources to which scientific men have gone in their attempts to reconstruct that primitive mentality are very various. Recently the science of psycho-analysis, which analyzes the way in which the egotistic and passionate impulses of the child are restrained, suppressed, modified or overlaid, to adapt them to the needs of social life, seems to have thrown a considerable amount of light upon the history of primitive society; and another fruitful source of suggestion has been the study of the ideas and customs of such contemporary savages as still survive. Again there is a sort of mental fossilization which we find in folk-lore and the deep-lying irrational superstitions and prejudices that still survive among modern civilized people. And finally we have in the increasingly numerous pictures, statues, carvings, symbols and the like, as we draw near to our own time, clearer and clearer indications of what man found interesting and worthy of record and representation.

Primitive man probably thought very much as a child thinks, that is to say in a series of imaginative pictures. He conjured up images or images presented themselves to his mind, and he acted in accordance with the emotions they aroused. So a child or an uneducated person does to-day. Systematic thinking is apparently a comparatively late development in human experience; it has not played any great part in human life until within the last three thousand years. And even to-day those who really control and order their thoughts are but a small minority of mankind. Most of the world still lives by imagination and passion.

Probably the earliest human societies, in the opening stages of the true human story, were small family groups. Just as the flocks and herds of the earlier mammals arose out of families which remained together and multiplied, so probably did the earliest tribes. But before this could happen a certain restraint upon the primitive egotisms of the individual had to be established. The fear of the father and respect for the mother had to be extended into adult life, and the natural jealousy of the old man of the group for the younger males as they grew up had to be mitigated. The mother on the other hand was the natural adviser and protector of the young. Human social life grew up out of the reaction between the crude instinct of the young to go off and pair by themselves as they grew up, on the one hand, and the dangers and disadvantages of separation on the other. An anthropological writer of great genius, J. J. Atkinson, in his *Primal Law*, has shown how much of the customary law of savages, the *Tabus*, that are so remarkable a fact in tribal life, can be ascribed to such a mental adjustment of the needs of the primitive human animal to a developing social life, and the later work of the psycho-analysts has done much to confirm his interpretation of these possibilities.

Some speculative writers would have us believe that respect and fear of the Old Man and the emotional reaction of the primitive savage to older protective women, exaggerated in dreams and enriched by fanciful mental play, played a large part in the beginnings of primitive religion and in the conception of gods and goddesses. Associated with this respect for powerful or helpful personalities was a dread and exaltation of such personages after their deaths, due to their reappearance in dreams. It was easy to believe they were not truly dead but only fantastically transferred to a remoteness of greater power.

The dreams, imaginations and fears of a child are far more vivid and real than those of a modern adult, and primitive man was always something of a child. He was nearer to the animals also, and he could suppose them to have motives and reactions like his own. He could imagine animal helpers, animal enemies, animal gods. One needs to have been an imaginative child oneself to realize again how important, significant, portentous or friendly, strangely shaped rocks, lumps of wood, exceptional trees or the like may have appeared to the men of the Old Stone Age, and how dream and fancy would create stories and legends about such

Brit. Mus.

RELICS OF THE STONE AGE

Chert implements from Somaliland. In general form they are similar to those found in Western and Northern Europe

things that would become credible as they told them. Some of these stories would be good enough to remember and tell again. The women would tell them to the children and so establish a tradition. To this day most imaginative children invent long stories in which some favourite doll or animal or some fantastic semi-human being figures as the hero, and primitive man probably did the same — with a much stronger disposition to believe his hero real.

For the very earliest of the true men that we know of were probably quite talkative beings. In that way they have differed from the Neanderthalers and had an advantage over them. The Neanderthaler may have been a dumb animal. Of course the primitive human speech was probably a very scanty collection of names, and may have been eked out with gestures and signs.

There is no sort of savage so low as not to have a kind of science of cause and effect. But primitive man was not very critical in his associations of cause with effect; he very easily connected an effect with something quite wrong as its cause. "You do so and so," he said, "and so and so happens." You give a child a poisonous berry and it dies. You eat the heart of a valiant enemy and you become strong. There we have two bits of cause and effect association, one true one false. We call the system of cause and effect in the mind of a savage, Fetish; but Fetish is simply savage science. It differs from modern science in that it is totally unsystematic and uncritical and so more frequently wrong.

In many cases it is not difficult to link cause and effect, in

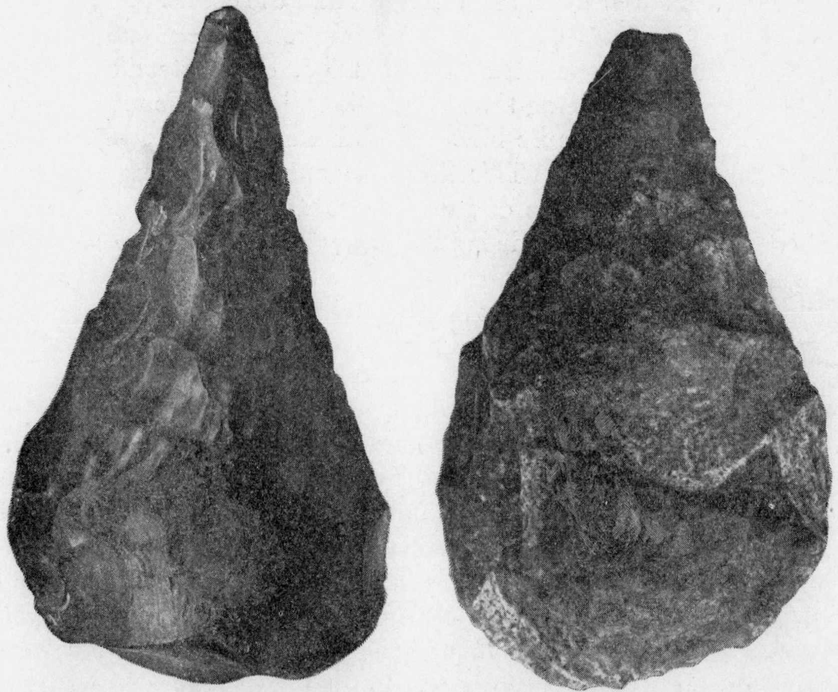

Brit. Mus.

WIDESPREAD SIMILARITY OF MEN OF THE STONE AGE

On the left is a flint implement excavated in Gray's Inn Lane, London; on the right one of similar form chipped by primitive men of Somaliland

many others erroneous ideas were soon corrected by experience; but there was a large series of issues of very great importance to primitive man, where he sought persistently for causes and found explanations that were wrong but not sufficiently wrong nor so obviously wrong as to be detected. It was a matter of great importance to him that game should be abundant or fish plentiful and easily caught, and no doubt he tried and believed in a thousand charms, incantations and omens to determine these desirable results. Another great concern of his was illness and death. Occasionally infections crept through the land and men died of them. Occasionally men were stricken by illness and died or were enfeebled without any manifest cause. This too must have given the hasty, emotional mind of primitive man much feverish exercise. Dreams and fantastic guesses made him blame this, or appeal for help to that man or beast or thing. He had the child's aptitude for fear and panic.

Quite early in the little human tribe, older, steadier minds sharing the fears, sharing the imaginations, but a little more forceful than the others, must have asserted themselves, to advise, to prescribe, to command. This they declared unpropitious and that imperative, this an omen of good and that an omen of evil. The expert in Fetish, the Medicine Man, was the first priest. He exhorted, he interpreted dreams, he warned, he performed the complicated hocus pocus that brought luck or averted calamity. Primitive religion was not so much what we now call religion as practice and observance, and the early priest dictated what was indeed an arbitrary primitive practical science.

XIII

The Beginnings of Cultivation

We are still very ignorant about the beginnings of cultivation and settlement in the world although a vast amount of research and speculation has been given to these matters in the last fifty years. All that we can say with any confidence at present is that somewhen about 15,000 and 12,000 B.C. while the Azilian people were in the south of Spain and while the remnants of the earlier hunters were drifting northward and eastward, somewhere in North Africa or Western Asia or in that great Mediterranean valley that is now submerged under the waters of the Mediterranean sea, there were people who, age by age, were working out two vitally important things: they were beginning cultivation and they were domesticating animals. They were also beginning to make, in addition to the chipped implements of their hunter forebears, implements of polished stone. They had discovered the possibility of basketwork and roughly woven textiles of plant fibre, and they were beginning to make a rudely modelled pottery.

They were entering upon a new phase in human culture, the Neolithic phase (New Stone Age) as distinguished from the Palæolithic (Old Stone) phase of the Cro-Magnards, the Grimaldi people, the Azilians and their like.[1] Slowly these Neolithic people spread over the warmer parts of the world; and the arts they had mastered, the plants and animals they had learnt to use, spread by imitation and acquisition even more widely than they did. By 10,000 B.C., most of mankind was at the Neolithic level.

[1] The term Palæolithic we may note is also used to cover the Neanderthaler and even the Eolithic implements. The pre-human age is called the "Older Palæolithic," the age of true men using unpolished stones in the "Newer Palæolithic."

Now the ploughing of land, the sowing of seed, the reaping of harvest, threshing and grinding, may seem the most obviously reasonable steps to a modern mind just as to a modern mind it is a commonplace that the world is round. What else could you do? people will ask. What else can it be? But to the primitive man of twenty thousand years ago neither of the systems of action and reasoning that seem so sure and manifest to us to-day were at all obvious. He felt his way to effectual practice through a multitude of trials and misconceptions, with fantastic and unnecessary elaborations and false interpretations at every turn. Somewhere in the Mediterranean region, wheat grew wild; and man may have learnt to pound and then grind up its seeds for food long before he learnt to sow. He reaped before he sowed.

And it is a very remarkable thing that throughout the world wherever there is sowing and harvesting there is still traceable the vestiges of a strong primitive association of the idea of sowing with the idea of a blood sacrifice, and primarily of the sacrifice of a human being. The study of the original entanglement of these two things is a profoundly attractive one to the curious mind; the interested reader will find it very fully developed in that monumental work, Sir J. G. Frazer's *Golden Bough.* It was an entanglement, we must remember, in the childish, dreaming, myth-making primitive mind; no reasoned process will explain it. But in that world of 12,000 to 20,000 years ago, it would seem that whenever seed time came round to the Neolithic peoples there was a human sacrifice. And it was not the sacrifice of any mean or outcast person; it was the sacrifice usually of a chosen youth or maiden, a youth more often who was treated with profound deference and even worship up to the moment of his immolation. He was a sort of sacrificial god-king, and all the details of his killing had become a ritual directed by the old, knowing men and sanctioned by the accumulated usage of ages.

At first primitive men, with only a very rough idea of the seasons, must have found great difficulty in determining when was the propitious moment for the seed-time sacrifice and the sowing. There is some reason for supposing that there was an early stage in human experience when men had no idea of a year. The first

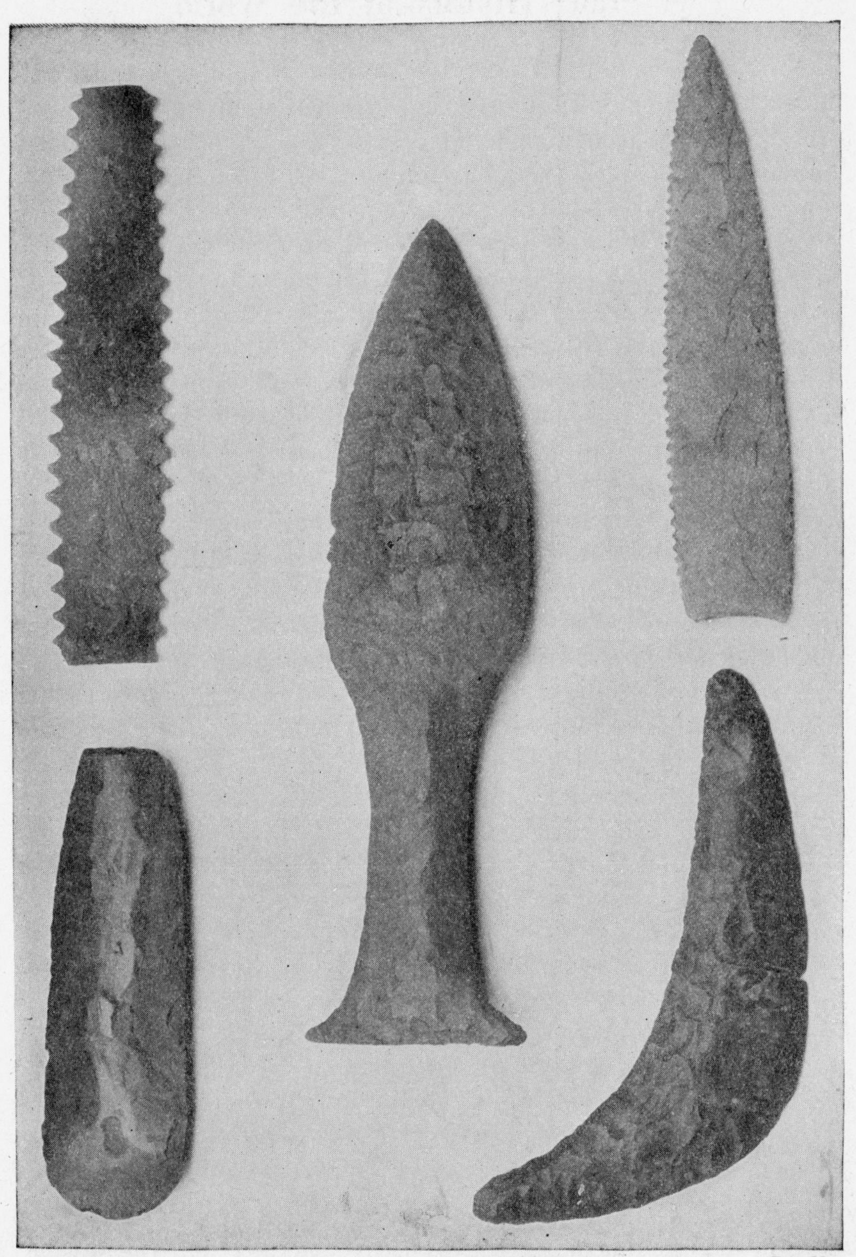

Brit. Mus.

NEOLITHIC FLINT IMPLEMENTS

1.

2.

Brit. Mus.

NEOLITHICISM OF TO-DAY

Spearheads, exactly as n the true Neolithic days, but made recently by Australian Natives.

(1) Made from a telegraph insulator;

(2) from a piece of broken bottle glass.

chronology was in lunar months; it is supposed that the years of the Biblical patriarchs are really moons, and the Babylonian calendar shows distinct traces of an attempt to reckon seed time by taking thirteen lunar months to see it round. This lunar influence upon the calendar reaches down to our own days. If usage did not dull our sense of its strangeness we should think it a very remarkable thing indeed that the Christian Church does not commemorate the Crucifixion and Resurrection of Christ on the proper anniversaries but on dates that vary year by year with the phases of the moon.

It may be doubted whether the first agriculturalists made any observation of the stars. It is more likely that stars were first observed by migratory herdsmen, who found them a convenient mark of direction. But once their use in determining seasons was realized, their importance to agriculture became very great. The seed-time sacrifice was linked up with the southing or northing of some prominent star. A myth and worship of that star was for primitive man an almost inevitable consequence.

It is easy to see how important the man of knowledge and experience, the man who knew about the blood sacrifice and the stars, became in this early Neolithic world.

The fear of uncleanness and pollution, and the methods of cleansing that were advisable, constituted another source of power for the knowledgeable men and women. For there have always been witches as well as wizards, and priestesses as well as priests. The early priest was really not so much a religious man as a man of applied science. His science was generally empirical and often bad; he kept it secret from the generality of men very jealously; but that does not alter the fact that his primary function was knowledge and that his primary use was a practical use.

Twelve or fifteen thousand years ago, in all the warm and fairly well-watered parts of the Old World these Neolithic human communities, with their class and tradition of priests and priestesses and their cultivated fields and their development of villages and little walled cities, were spreading. Age by age a drift and exchange of ideas went on between these communities. Eliot Smith

Brit. Mus.
SPECIMEN OF NEOLITHIC POTTERY
Dug up at Mortlake from the Thames Bed

and Rivers have used the term "Heliolithic culture" for the culture of these first agricultural peoples. "Heliolithic" (Sun and Stone) is not perhaps the best possible word to use for this, but until scientific men give us a better one we shall have to use it. Originating somewhere in the Mediterranean and western Asiatic area, it spread age by age eastward and from island to island across the Pacific until it may even have reached America and mingled with the more primitive ways of living of the Mongoloid immigrants coming down from the North.

Wherever the brownish people with the Heliolithic culture went they took with them all or most of a certain group of curious ideas and practices. Some of them are such queer ideas that they call for the explanation of the mental expert. They made pyramids and great mounds, and set up great circles of big stones, perhaps to facilitate the astronomical observation of the priests; they made mummies of some or all of their dead; they tattooed and circumcized; they had the old custom, known as the *couvade*, of sending the *father* to bed and rest when a child was born, and they had as a luck symbol the well-known Swastika.

If we were to make a map of the world with dots to show how far these group practices have left their traces, we should make a belt along the temperate and sub-tropical coasts of the world from Stonehenge and Spain across the world to Mexico and Peru. But Africa below the equator, north central Europe, and north Asia would show none of these dottings; there lived races who were developing along practically independent lines.

XIV

Primitive Neolithic Civilizations

ABOUT 10,000 B.C. the geography of the world was very similar in its general outline to that of the world to-day. It is probable that by that time the great barrier across the Straits of Gibraltar that had hitherto banked back the ocean waters from the Mediterranean valley had been eaten through, and that the Mediterranean was a sea following much the same coastlines as it does now. The Caspian Sea was probably still far more extensive than it is at present, and it may have been continuous with the Black Sea to the north of the Caucasus Mountains. About this great Central Asian sea lands that are now steppes and deserts were fertile and habitable. Generally it was a moister and more fertile world. European Russia was much more a land of swamp and lake than it is now, and there may still have been a land connexion between Asia and America at Behring Straits.

It would have been already possible at that time to have distinguished the main racial divisions of mankind as we know them to-day. Across the warm temperate regions of this rather warmer and better-wooded world, and along the coasts, stretched the brownish peoples of the Heliolithic culture, the ancestors of the bulk of the living inhabitants of the Mediterranean world, of the Berbers, the Egyptians and of much of the population of South and Eastern Asia. This great race had of course a number of varieties. The Iberian or Mediterranean or "dark-white" race of the Atlantic and Mediterranean coast, the "Hamitic" peoples which include the Berbers and Egyptians, the Dravidians, the darker people of India, a multitude of East Indian people, many Polynesian races and the Maoris are all divisions of various value of this great main mass of humanity. Its western varieties are whiter than its eastern. In the forests of central and northern Europe a more blonde variety of men with blue eyes was becoming distinguishable, branching off from the main mass of brownish people, a variety which many people now speak of as the Nordic race. In the more open regions of north-eastern Asia was another differentiation of this brownish humanity in the direction of a type with more oblique eyes, high cheek-bones, a yellowish skin, and very straight black hair, the Mongolian peoples. In South Africa, Australia, in many tropical islands in the south of Asia were remains of the early negroid peoples. The central parts of Africa were already a region of racial intermixture. Nearly all the coloured races of Africa to-day seem to be blends of the brownish peoples of the north with a negroid substratum.

We have to remember that human races can all interbreed freely and that they separate, mingle and reunite as clouds do. Human races do not branch out like trees with branches that never come together again. It is a thing we need to bear constantly in mind, this remingling of races at any opportunity. It will save us from many cruel delusions and prejudices if we do so. People will use such a word as race in the loosest manner, and base the most preposterous generalizations upon it. They will speak of a "British"

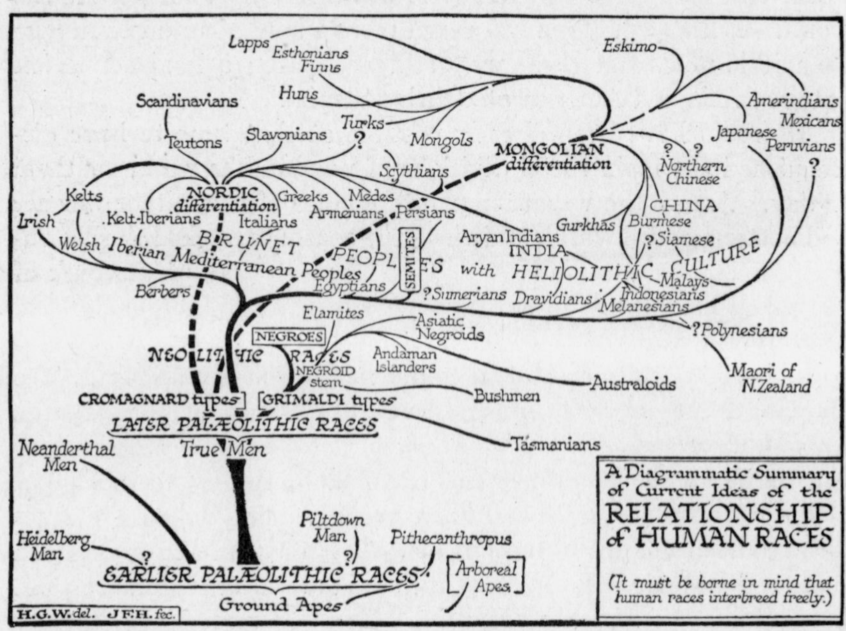

A MAYA STELE Brit. Mus.

Showing a worshipper and a Serpent God. Note the grotesque faces in the writing

race or of a "European" race. But nearly all the European nations are confused mixtures of brownish, dark-white, white and Mongolian elements.

It was at the Neolithic phase of human development that peoples of the Mongolian breed first made their way into America. Apparently they came by way of Behring Straits and spread southward. They found caribou, the American reindeer, in the north and great herds of bison in the south. When they reached South America there were still living the Glyptodon, a gigantic armadillo, and the Megatherium, a monstrous clumsy sloth as high as an elephant. They probably exterminated the latter beast, which was as helpless as it was big.

The greater portion of these American tribes never rose above a hunting nomadic Neolithic life. They never discovered the use of iron, and their chief metal possessions were native gold and copper. But in Mexico, Yucatan and Peru conditions existed favourable to settled cultivation, and here about 1000 B.C. or so arose very interesting civilizations of a parallel but different type from the old-world civilization. Like the much earlier primitive civilizations of the old world these communities displayed a great development of human sacrifice about the processes of seed time and harvest; but while in the old world, as we shall see, these primary ideas were ultimately mitigated, complicated and overlaid by others, in America they developed and were elaborated to a very high degree of intensity. These American civilized countries were essentially priest-ruled countries; their war chiefs and rulers were under a rigorous rule of law and omen.

These priests carried astronomical science to a high level of accuracy. They knew their year better than the Babylonians of whom we shall presently tell. In Yucatan they had a kind of writing, the Maya writing, of the most curious and elaborate character. So far as we have been able to decipher it, it was used mainly for keeping the exact and complicated calendars upon which the priests expended their intelligence. The art of the Maya civilization came to a climax about 700 or 800 A.D. The sculptured work of these people amazes the modern observer by its great plastic power and its frequent beauty, and perplexes him by a grotesqueness and by a sort of insane conventionality and intricacy outside the circle of his ideas. There is nothing quite like it in the old world. The nearest approach, and that is a remote one, is found in archaic Indian carvings. Everywhere there are woven feathers and serpents twine in and out. Many Maya inscriptions resemble a certain sort of elaborate drawing made by lunatics in European asylums, more than any other old-world work. It is as if the Maya mind had developed upon a different line from the old-world mind, had a different twist to its ideas, was not, by old-world standards, a rational mind at all.

This linking of these aberrant American civilizations to the idea of a general mental aberration finds support in their extraordinary obsession by the shedding of human blood. The Mexican civilization in particular ran blood; it offered thousands of human victims yearly. The cutting open of living victims, the tearing out of the still beating heart, was an act that dominated the minds and lives of these strange priesthoods. The public life, the national festivities all turned on this fantastically horrible act.

NEOLITHIC WARRIOR

Modelled from drawing by Prof. Rutot

The ordinary existence of the common people in these communities was very like the ordinary existence of any other barbaric peasantry. Their pottery, weaving and dyeing was very good. The Maya writing was not only carven on stone but written and painted upon skins and the like. The European and American museums contain many enigmatical Maya manuscripts of which at present little has been deciphered except the dates. In Peru there were beginnings of a similar writing but they were superseded by a method of keeping records by knotting cords. A similar method of mnemonics was in use in China thousands of years ago.

In the old world before 4000 or 5000 B.C., that is to say three or four thousand years earlier, there were primitive civilizations not unlike these American civilizations; civilizations based upon a temple, having a vast quantity of blood sacrifices and with an intensely astronomical priesthood. But in the old world the primitive civilizations reacted upon one another and developed towards the conditions of our own world. In America these primitive civilizations never progressed beyond this primitive stage. Each of them was in a little world of its own. Mexico it seems knew little or nothing of Peru, until the Europeans came to America. The potato, which was the principal food stuff in Peru, was unknown in Mexico.

Age by age these peoples lived and marvelled at their gods and made their sacrifices and died. Maya art rose to high levels of decorative beauty. Men made love and tribes made war. Drought and plenty, pestilence and health, followed one another. The priests elaborated their calendar and their sacrificial ritual through long centuries, but made little progress in other directions.

XV

Sumeria, Early Egypt and Writing

THE old world is a wider, more varied stage than the new. By 6000 or 7000 B.C. there were already quasi-civilized communities almost at the Peruvian level, appearing in various fertile regions of Asia and in the Nile valley. At that time north Persia and western Turkestan and south Arabia were all more fertile than they are now, and there are traces of very early communities in these regions. It is in lower Mesopotamia however and in Egypt that there first appear cities, temples, systematic irrigation, and evidences of a social organization rising above the level of a mere barbaric village-town. In those days the Euphrates and Tigris flowed by separate mouths into the Persian Gulf, and it was in the country between them that the Sumerians built their first cities. About the same time, for chronology is still vague, the great history of Egypt was beginning.

These Sumerians appear to have been a brownish people with prominent noses. They employed a sort of writing that has been deciphered, and their language is now known. They had discovered the use of bronze and they built great tower-like temples of sun-dried brick. The clay of this country is very fine; they used it to write upon, and so it is that their inscriptions have been preserved to us. They had cattle, sheep, goats and asses, but no horses. They fought on foot, in close formation, carrying spears and shields of skin. Their clothing was of wool and they shaved their heads.

Each of the Sumerian cities seems generally to have been an independent state with a god of its own and priests of its own. But sometimes one city would establish an ascendancy over others and exact tribute from their population. A very ancient inscription at Nippur records the "empire," the first recorded empire, of the Sumerian city of Erech. Its god and its priest-king claimed an authority from the Persian Gulf to the Red Sea.

At first writing was merely an abbreviated method of pictorial record. Even before Neolithic times men were beginning to write. The Azilian rock pictures to which we have already referred show the beginning of the process. Many of them record hunts and expeditions, and in most of these the human figures are plainly drawn. But in some the painter would not bother with head and limbs; he just indicated men by a vertical and one or two transverse strokes. From this to a conventional condensed picture writing was an easy transition. In Sumeria, where the writing was done on clay with a stick, the dabs of the characters soon became unrecognizably unlike the things they stood for, but in Egypt where men painted on walls and on strips of the papyrus reed (the first paper) the likeness to the thing imitated remained. From the fact that the wooden styles used in Sumeria made wedge-shaped marks, the Sumerian writing is called cuneiform (= wedge-shaped).

BRICK OF HAMMURABI, KING OF BABYLON ABOUT 2200 B.C.
Note the cuneiform characters of the inscription, which records the building of a temple to a Sun God

An important step towards writing was made when pictures were used to indicate not the thing represented but some similar thing. In the rebus dear to children of a suitable age, this is still done to-day. We draw a camp with tents and a bell, and the child is delighted to guess that this is the Scotch name Campbell. The Sumerian language was a language made up of accumulated syllables rather like some contemporary Amerindian languages, and it lent itself very readily to this syllabic method of writing words expressing ideas that could not be conveyed by pictures directly. Egyptian writing underwent parallel developments. Later on, when foreign peoples with less distinctly syllabled methods of speech were to learn and use these picture scripts they were to make those further modifications and simplifications that developed at last into alphabetical writing. All the true alphabets of the later world derived from a mixture of the Sumerian cuneiform and the Egyptian hieroglyphic (priest writing). Later in China there was to develop a conventionalized picture writing, but in China it never got to the alphabetical stage.

EBONY CYLINDER SEALS OF FIRST EGYPTIAN DYNASTY

Recovered from the Tombs at Abydos in 1921 by the British School of Archæology. They give evidence of early form of block printing

The invention of writing was of very great importance in the development of human societies. It put agreements, laws, commandments on record. It made the growth of states larger than the old city states possible. It made a continuous historical consciousness possible. The command of the priest or king and his seal could go far beyond his sight and voice and could survive his death. It is interesting to note that in ancient Sumeria seals were greatly used.

Photo: J. Boyer

THE SAKHARA PYRAMIDS

The Pyramid to right, the step Pyramid, is the oldest stone building in the world

A king or a nobleman or a merchant would have his seal often very artistically carved, and would impress it on any clay document he wished to authorize. So close had civilization got to printing six thousand years ago. Then the clay was dried hard and became permanent. For the reader must remember that in the land of Mesopotamia for countless years, letters, records and accounts were all written on comparatively indestructible tiles. To that fact we owe a great wealth of recovered knowledge.

Bronze, copper, gold, silver and, as a precious rarity, meteoric iron were known in both Sumeria and Egypt at a very early stage.

Daily life in those first city lands of the old world must have been

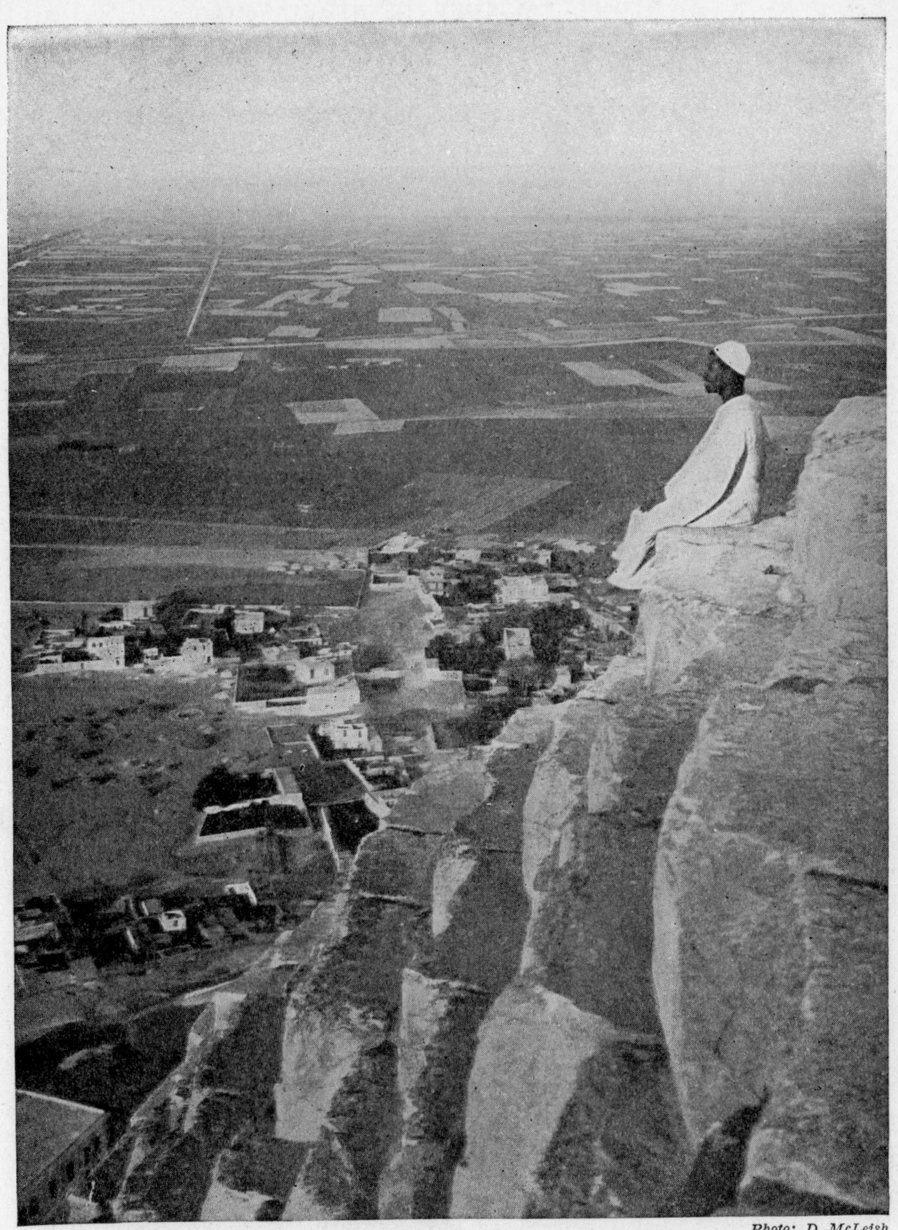

Photo: D. McLeish

VIEW FROM THE SUMMIT OF THE GREAT PYRAMID OF CHEOPS

Showing how these great monuments dominate the plain

Photo: D. McLeish

THE TEMPLE OF HATHOR AT DENDEREH

very similar in both Egypt and Sumeria. And except for the asses and cattle in the streets it must have been not unlike the life in the Maya cities of America three or four thousand years later. Most of the people in peace time were busy with irrigation and cultivation — except on days of religious festivity. They had no money and no need for it. They managed their small occasional trades by barter. The princes and rulers who alone had more than a few possessions used gold and silver bars and precious stones for any incidental act of trade. The temple dominated life; in Sumeria it was a great towering temple that went up to a roof from which the stars were observed; in Egypt it was a massive building with only a ground floor. In Sumeria the priest ruler was the greatest, most splendid of beings. In Egypt however there was one who was raised above the priests; he was the living incarnation of the chief god of the land, the Pharaoh, the god king.

There were few changes in the world in those days; men's days were sunny, toilsome and conventional. Few strangers came into the land and such as did fared uncomfortably. The priest directed life according to immemorial rules and watched the stars for seed time and marked the omens of the sacrifices and interpreted the warnings of dreams. Men worked and loved and died, not unhappily, forgetful of the savage past of their race and heedless of its future. Sometimes the ruler was benign. Such was Pepi II, who reigned in Egypt for ninety years. Sometimes he was ambitious and took men's sons to be soldiers and sent them against neighbouring city states to war and plunder, or he made them toil to build great buildings. Such were Cheops and Chephren and Mycerinus, who built those vast sepulchral piles, the pyramids at Gizeh. The largest of these is 450 feet high and the weight of stone in it is 4,883,000 tons. All this was brought down the Nile in boats and lugged into place chiefly by human muscle. Its erection must have exhausted Egypt more than a great war would have done.

XVI

Primitive Nomadic Peoples

IT was not only in Mesopotamia and the Nile Valley that men were settling down to agriculture and the formation of city states in the centuries between 6000 and 3000 B.C. Wherever there were possibilities of irrigation and a steady all-the-year-round food supply men were exchanging the uncertainties and hardships of hunting and wandering for the routines of settlement. On the upper Tigris a people called the Assyrians were founding cities; in the valleys of Asia Minor and on the Mediterranean shores and islands, there were small communities growing up to civilization. Possibly parallel developments of human life were already going on in favourable regions of India and China. In many parts of Europe where there were lakes well stocked with fish, little communities of men had long settled in dwellings built on piles over the water, and were eking out agriculture by fishing and hunting. But over much larger areas of the old world no such settlement was possible. The land was too harsh, too thickly wooded or too arid, or the seasons too uncertain for mankind, with only the implements and science of that age to take root.

For settlement under the conditions of the primitive civilizations men needed a constant water supply and warmth and sunshine. Where these needs were not satisfied, man could live as a transient, as a hunter following his game, as a herdsman following the seasonal grass, but he could not settle. The transition from the hunting to the herding life may have been very gradual. From following herds of wild cattle or (in Asia) wild horses, men may have come to an idea of property in them, have learnt to pen them into valleys, have fought for them against wolves, wild dogs and other predatory beasts.

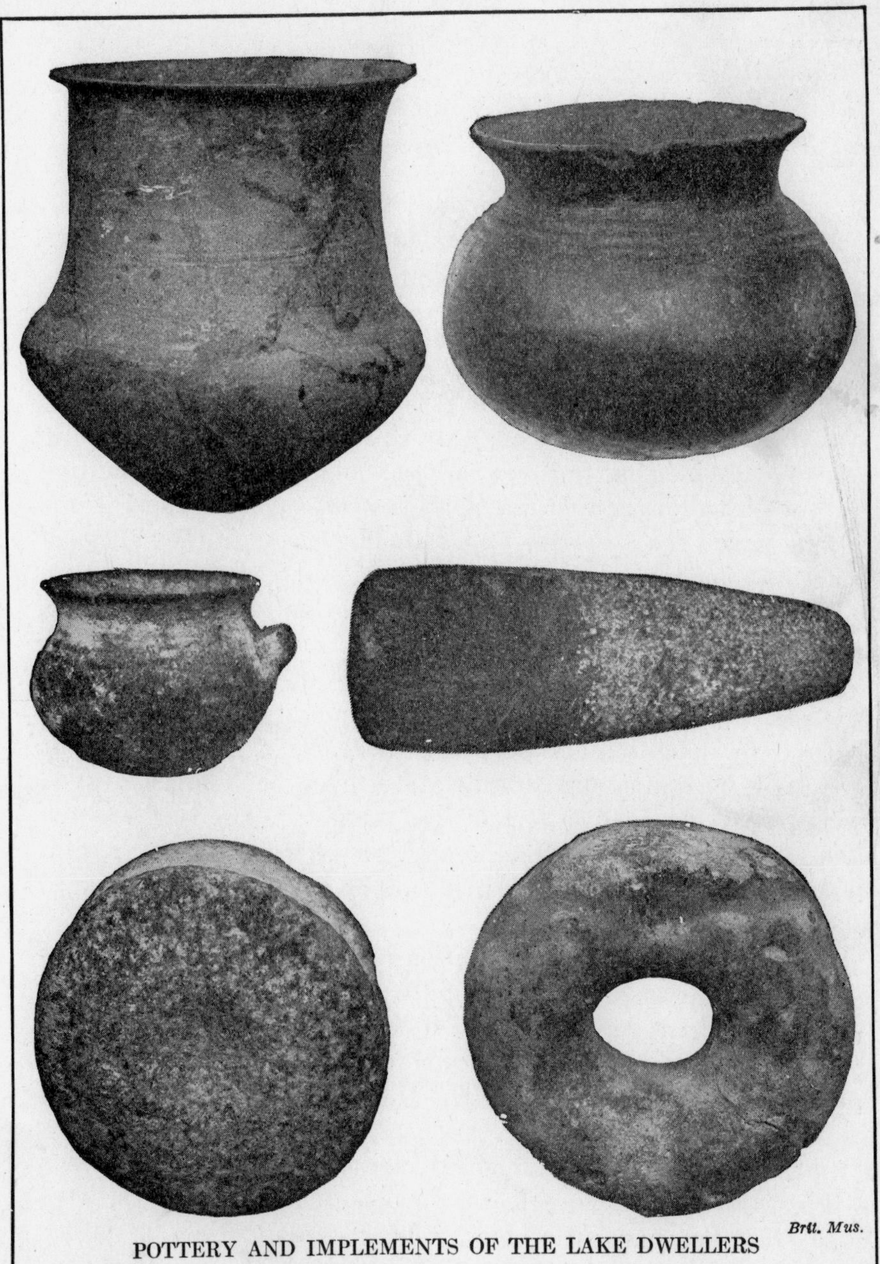

POTTERY AND IMPLEMENTS OF THE LAKE DWELLERS

A CONTEMPORARY LAKE VILLAGE

These Borneo dwellings are practically counterparts of the homes of European neolithic communities 6000 B.C.

So while the primitive civilizations of the cultivators were growing up chiefly in the great river valleys, a different way of living, the nomadic life, a life in constant movement to and fro from winter pasture to summer pasture, was also growing up. The nomadic peoples were on the whole hardier than the agriculturalists; they were less prolific and numerous, they had no permanent temples and no highly organized priesthood; they had less gear; but the reader must not suppose that theirs was necessarily a less highly developed way of living on that account. In many ways this free life was a fuller life than that of the tillers of the soil. The individual was more self-reliant; less of a unit in a crowd. The leader was more important; the medicine man perhaps less so.

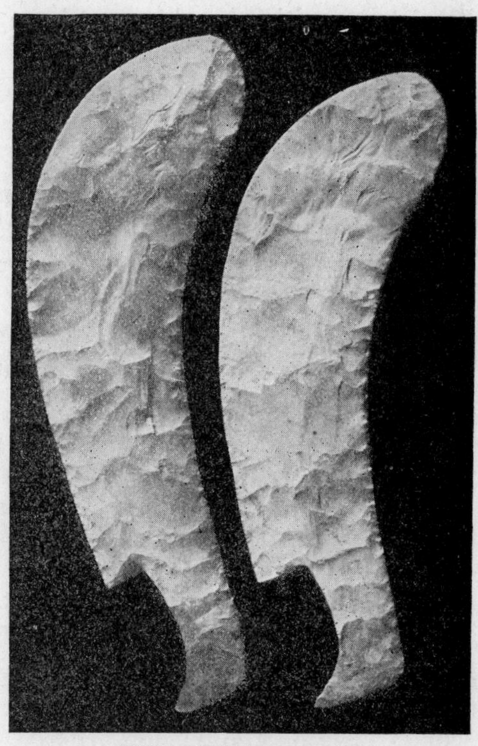

FLINT KNIVES OF 4500 B.C.
Excavated 1922 by the British School of Archæology in Egypt from First Dynasty Tombs

Moving over large stretches of country the nomad took a wider view of life. He touched on the confines of this settled land and that. He was used to the sight of strange faces. He had to scheme and treat for pasture with competing tribes. He knew more of minerals than the folk upon the plough lands because he went over mountain passes and into rocky places. He may have been a better metallurgist. Possibly bronze and much more probably iron smelting were nomadic discoveries. Some of the earliest implements of iron reduced from its ores have been found in Central Europe far away from the early civilizations.

On the other hand the settled folk had their textiles and their pottery and made many desirable things. It was inevitable that as the two sorts of life, the agricultural and the nomadic differentiated, a certain amount of looting and trading should develop between the two. In Sumeria particularly which had deserts and seasonal

NOMADS IN EGYPT

Egyptian wall painting in a tomb near ancient Beni Hassan, middle Egypt. It depicts the arrival of a tribe of Semitic Nomads in Egypt about the year 1895 B.C

country on either hand it must have been usual to have the nomads camping close to the cultivated fields, trading and stealing and perhaps tinkering, as gipsies do to this day. (But hens they would not steal, because the domestic fowl — an Indian jungle fowl originally — was not domesticated by man until about 1000 B.C.) They would bring precious stones and things of metal and leather. If they were hunters they would bring skins. They would get in exchange pottery and beads and glass, garments and suchlike manufactured things.

Three main regions and three main kinds of wandering and imperfectly settled people there were in those remote days of the first civilizations in Sumeria and early Egypt. Away in the forests of Europe were the blonde Nordic peoples, hunters and herdsmen, a lowly race. The primitive civilizations saw very little of this race before 1500 B.C. Away on the steppes of eastern Asia various Mongolian tribes, the Hunnish peoples, were domesticating the horse and developing a very wide sweeping habit of seasonal movement between their summer and winter camping places. Possibly the Nordic and Hunnish peoples were still separated from one another by the swamps of Russia and the greater Caspian Sea of that time. For very much of Russia there was swamp and lake. In the deserts, which were growing more arid now, of Syria and Arabia, tribes of a dark white or brownish people, the Semitic tribes, were driving flocks of sheep and goats and asses from pasture to pasture. It was these Semitic shepherds and certain more negroid people from southern Persia, the Elamites, who were the first nomads to come into close contact with the early civilizations. They came

EGYPT PEASANTS GOING TO WORK

From an ancient and curiously painted model in the British Museum

STELE GLORIFYING KING NARAM SIN, OF AKKAD

This monarch, son of Sargon I, was a great architect as well as a famous conqueror. Discovered in 1898 among the ruins of Susa, Persia

as traders and as raiders. Finally there arose leaders among them with bolder imaginations, and they became conquerors.

About 2750 B.C. a great Semitic leader, Sargon, had conquered the whole Sumerian land and was master of all the world from the Persian Gulf to the Mediterranean Sea. He was an illiterate barbarian and his people, the Akkadians, learnt the Sumerian writing and adopted the Sumerian language as the speech of the officials and the learned. The empire he founded decayed after two centuries, and after one inundation of Elamites a fresh Semitic people, the Amorites, by degrees established their rule over Sumeria. They made their capital in what had hitherto been a small up-river town, Babylon, and their empire is called the first Babylonian Empire. It was consolidated by a great king called Hammurabi (circa 2100 B.C.) who made the earliest code of laws yet known to history.

The narrow valley of the Nile lies less open to nomadic invasion than Mesopotamia, but about the time of Hammurabi occurred a successful Semitic invasion of Egypt and a line of Pharaohs was set up, the Hyksos or "shepherd kings," which lasted for several centuries. These Semitic conquerors never assimilated themselves with the Egyptians; they were always regarded with hostility as foreigners and barbarians; and they were at last expelled by a popular uprising about 1600 B.C.

But the Semites had come into Sumeria for good and all, the two races assimilated and the Babylonian Empire became Semitic in its language and character.

XVII

The First Sea-going Peoples

THE earliest boats and ships must have come into use some twenty-five or thirty thousand years ago. Man was probably paddling about on the water with a log of wood or an inflated skin to assist him, at latest in the beginnings of the Neolithic period. A basketwork boat covered with skin and caulked was used in Egypt and Sumeria from the beginnings of our knowledge. Such boats are still used there. They are used to this day in Ireland and Wales and in Alaska; sealskin boats still make the crossing of Behring Straits. The hollow log followed as tools improved. The building of boats and then ships came in a natural succession.

Perhaps the legend of Noah's Ark preserves the memory of some early exploit in shipbuilding, just as the story of the Flood, so widely distributed among the peoples of the world, may be the tradition of the flooding of the Mediterranean basin.

There were ships upon the Red Sea long before the pyramids were built, and there were ships on the Mediterranean and Persian Gulf by 7000 B.C. Mostly these were the ships of fishermen, but some were already trading and pirate ships — for knowing what we do of mankind we may guess pretty safely that the first sailors plundered where they could and traded where they had to do so.

The seas on which these first ships adventured were inland seas on which the wind blew fitfully and which were often at a dead calm for days together, so that sailing did not develop beyond an accessory use. It is only in the last four hundred years that the well-rigged, ocean-going, sailing ship has developed. The ships of the ancient world were essentially rowing ships which hugged the shore and went into harbour at the first sign of rough weather. As ships grew into big galleys they caused a demand for war captives as galley slaves.

We have already noted the appearance of the Semitic people as wanderers and nomads in the region of Syria and Arabia, and how they conquered Sumeria and set up first the Akkadian and then the first Babylonian Empire. In the west these same Semitic peoples were taking to the sea. They set up a string of harbour towns along the Eastern coast of the Mediterranean, of which Tyre and Sidon were the chief; and by the time of Hammurabi in Babylon, they had spread as traders, wanderers and colonizers over the whole Mediterranean basin. These sea Semites were called the Phœnicians. They settled largely in Spain, pushing back the old Iberian Basque population and sending coasting expeditions through the straits of Gibraltar; and they set up colonies upon the north coast of Africa. Of Carthage, one of these Phœnician cities, we shall have much more to tell later.

But the Phœnicians were not the first people to have galleys in the Mediterranean waters. There was already a series of towns and cities among the islands and coasts of that sea belonging to a race or races apparently connected by blood and language with the Basques to the west and the Berbers and Egyptians to the south, the Ægean peoples. These peoples must not be confused with the Greeks, who come much later into our story; they were pre-Greek, but they had cities in Greece and Asia Minor, Mycenæ and Troy for example, and they had a great and prosperous establishment at Cnossos in Crete.

It is only in the last half century that the industry of excavating archæologists has brought the extent and civilization of the Ægean peoples to our knowledge. Cnossos has been most thoroughly explored; it was happily not succeeded by any city big enough to destroy its ruins, and so it is our chief source of information about this once almost forgotten civilization.

The history of Cnossos goes back as far as the history of Egypt; the two countries were trading actively across the sea by 4000 B.C. By 2500 B.C., that is between the time of Sargon I and Hammurabi, Cretan civilization was at its zenith.

Cnossos was not so much a town as a great palace for the Cretan monarch and his people. It was not even fortified. It was only fortified later as the Phœnicians grew strong, and as a new and more terrible breed of pirates, the Greeks, came upon the sea from the north.

The monarch was called Minos, as the Egyptian monarch was called Pharaoh; and he kept his state in a palace fitted with running water, with bathrooms and the like conveniences such as we know of in no other ancient remains. There he held great festivals and shows. There was bull-fighting, singularly like the bull-fighting that still survives in Spain; there was resemblance even in the costumes of the bull-fighters; and there were gymnastic displays. The

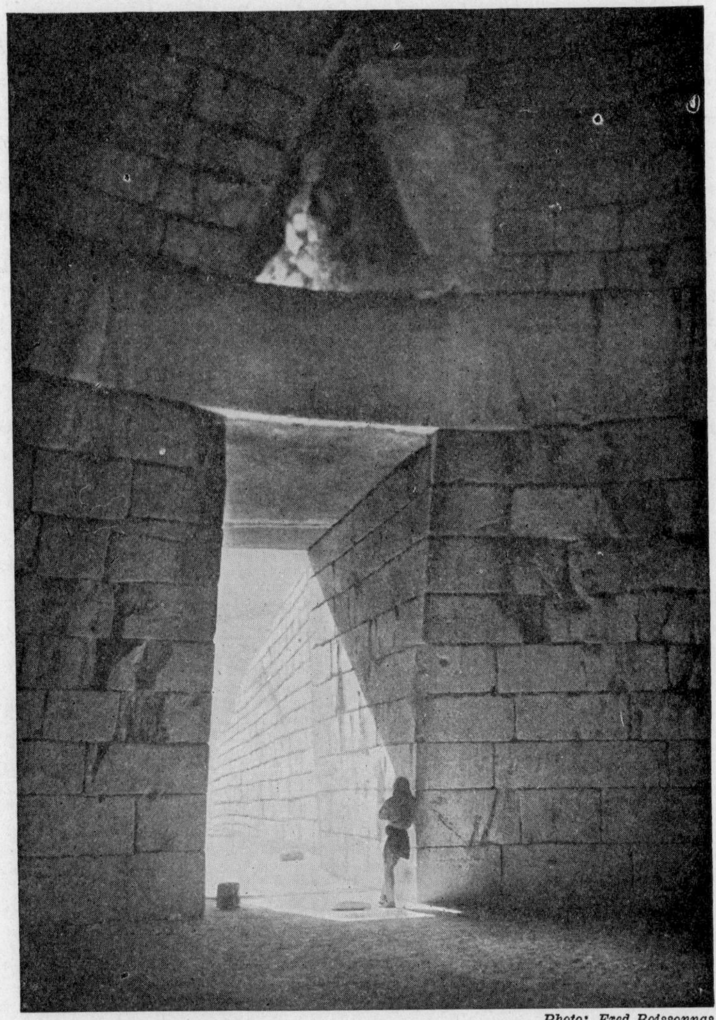

Photo: Fred Boissonnas

THE TREASURE HOUSE AT MYCENÆ

women's clothes were remarkably modern in spirit; they wore corsets and flounced dresses. The pottery, the textile manufactures, the sculpture, painting, jewellery, ivory, metal and inlay work of these

Cretans was often astonishingly beautiful. And they had a system of writing, but that still remains to be deciphered.

This happy and sunny and civilized life lasted for some score of centuries. About 2000 B.C. Cnossos and Babylon abounded in comfortable and cultivated people who probably led very pleasant lives. They had shows and they had religious festivals, they had domestic slaves to look after them and industrial slaves to make a profit for them. Life must have seemed very secure in Cnossos for such people, sunlit and girdled by the blue sea. Egypt of course must have appeared rather a declining country in those days under the rule of her half-barbaric shepherd kings, and if one took an interest in politics one must have noticed how the Semitic people seemed to be getting everywhere, ruling Egypt, ruling distant Babylon, building Nineveh on the upper Tigris, sailing west to the Pillars of Hercules (the straits of Gibraltar) and setting up their colonies on those distant coasts.

There were some active and curious minds in Cnossos, because later on the Greeks told legends of a certain skilful Cretan artificer, Dædalus, who attempted to make some sort of flying machine, perhaps a glider, which collapsed and fell into the sea.

It is interesting to note some of the differences as well as the resemblances between the life of Cnossos and our own. To a Cretan gentleman of 2500 B.C. iron was a rare metal which fell out of the sky and was curious rather than useful—for as yet only meteoric iron was known, iron had not been obtained from its ores. Compare that with our modern state of affairs pervaded by iron everywhere. The horse again would be a quite legendary creature to our Cretan, a sort of super-ass which lived in the bleak northern lands far away beyond the Black Sea. Civilization for him dwelt chiefly in Ægean Greece and Asia Minor, where Lydians and Carians and Trojans lived a life and probably spoke languages like his own. There were Phœnicians and Ægeans settled in Spain and North Africa, but those were very remote regions to his imagination. Italy was still a desolate land covered with dense forests; the brown-skinned Etruscans had not yet gone there from Asia Minor. And one day perhaps this Cretan gentleman went down to the harbour and saw a captive who attracted his attention because he was very fair-complexioned and had blue eyes. Perhaps our Cretan tried to talk to him and was answered in an unintelligible gibberish. This creature came from somewhere beyond the Black Sea and seemed to be an altogether benighted savage. But indeed he was an Aryan tribesman, of a race and culture of which we shall soon have much to tell, and the strange gibberish he spoke was to differentiate some day into Sanskrit,

Photo: *Fred Boissonnas*

THE PALACE AT CNOSSOS
The painted walls of the Throne Room

Persian, Greek, Latin, German, English and most of the chief languages of the world.

Such was Cnossos at its zenith, intelligent, enterprising, bright and happy. But about 1400 B.C. disaster came perhaps very suddenly upon its prosperity. The palace of Minos was destroyed, and its ruins have never been rebuilt or inhabited from that day to this. We do not know how this disaster occurred. The excavators note what appears to be scattered plunder and the marks of the fire. But the traces of a very destructive earthquake have also been found. Nature alone may have destroyed Cnossos, or the Greeks may have finished what the earthquake began.

XVIII

Egypt, Babylon and Assyria

THE Egyptians had never submitted very willingly to the rule of their Semitic shepherd kings and about 1600 A.D. a vigorous patriotic movement expelled these foreigners. Followed a new phase or revival for Egypt, a period known to Egyptologists as the New Empire. Egypt, which had not been closely consolidated before the Hyksos invasion, was now a united country; and the phase of subjugation and insurrection left her full of military spirit. The Pharaohs became aggressive conquerors. They had now acquired the war horse and the war chariot, which the Hyksos had brought to them. Under Thothmes III and Amenophis III Egypt had extended her rule into Asia as far as the Euphrates.

We are entering now upon a thousand years of warfare between the once quite separated civilizations of Mesopotamia and the Nile. At first Egypt was ascendant. The great dynasties, the Seventeenth Dynasty, which included Thothmes III and Amenophis III and IV and a great queen Hatasu, and the Nineteenth, when Rameses II, supposed by some to have been the Pharaoh of Moses, reigned for sixty-seven years, raised Egypt to high levels of prosperity. In between there were phases of depression for Egypt, conquest by the Syrians and later conquest by the Ethiopians from the South. In Mesopotamia Babylon ruled, then the Hittites and the Syrians of Damascus rose to a transitory predominance; at one time the Syrians conquered Egypt; the fortunes of the Assyrians of Nineveh ebbed and flowed; sometimes the city was a conquered city; sometimes the Assyrians ruled in Babylon and assailed Egypt. Our space is too limited here to tell of the comings and goings of the armies of the Egyptians and of the various Semitic powers of Asia Minor, Syria and Mesopotamia. They were armies now provided with vast droves of war chariots, for the horse — still used only for war and glory — had spread by this time into the old civilizations from Central Asia.

Great conquerors appear in the dim light of that distant time and pass, Tushratta, King of Mitanni, who captured Nineveh, Tiglath Pileser I of Assyria who conquered Babylon. At last the Assyrians became the greatest military power of the time. Tiglath

TEMPLE AT ABU SIMBEL
Showing statues of Rameses II at entrance

Pileser III conquered Babylon in 745 B.C. and founded what historians call the New Assyrian Empire. Iron had also come now into civilization out of the north; the Hittites, the precursors of the Armenians, had it first and communicated its use to the Assyrians, and an Assyrian usurper, Sargon II, armed his troops with it. Assyria became the first power to expound the doctrine of blood and iron. Sargon's son Sennacherib led an army to the borders of Egypt, and was defeated not by military strength but by the plague. Sennacherib's grandson Assurbanipal (who is also known in history by his Greek name of Sardanapalus) did actually conquer Egypt in 670 B.C. But Egypt was already a conquered country then under an Ethiopian dynasty. Sardanapalus simply replaced one conqueror by another.

If one had a series of political maps of this long period of history, this interval of ten centuries, we should have Egypt expanding and

Photo: D. McLeish

AVENUE OF SPHINXES
Leading from the Nile to the great Temple of Karnak

contracting like an amœba under a microscope, and we should see these various Semitic states of the Babylonians, the Assyrians, the Hittites and the Syrians coming and going, eating each other up and disgorging each other again. To the west of Asia Minor there would be little Ægean states like Lydia, whose capital was Sardis, and Caria. But after about 1200 B.C. and perhaps earlier, a new set of names would come into the map of the ancient world from

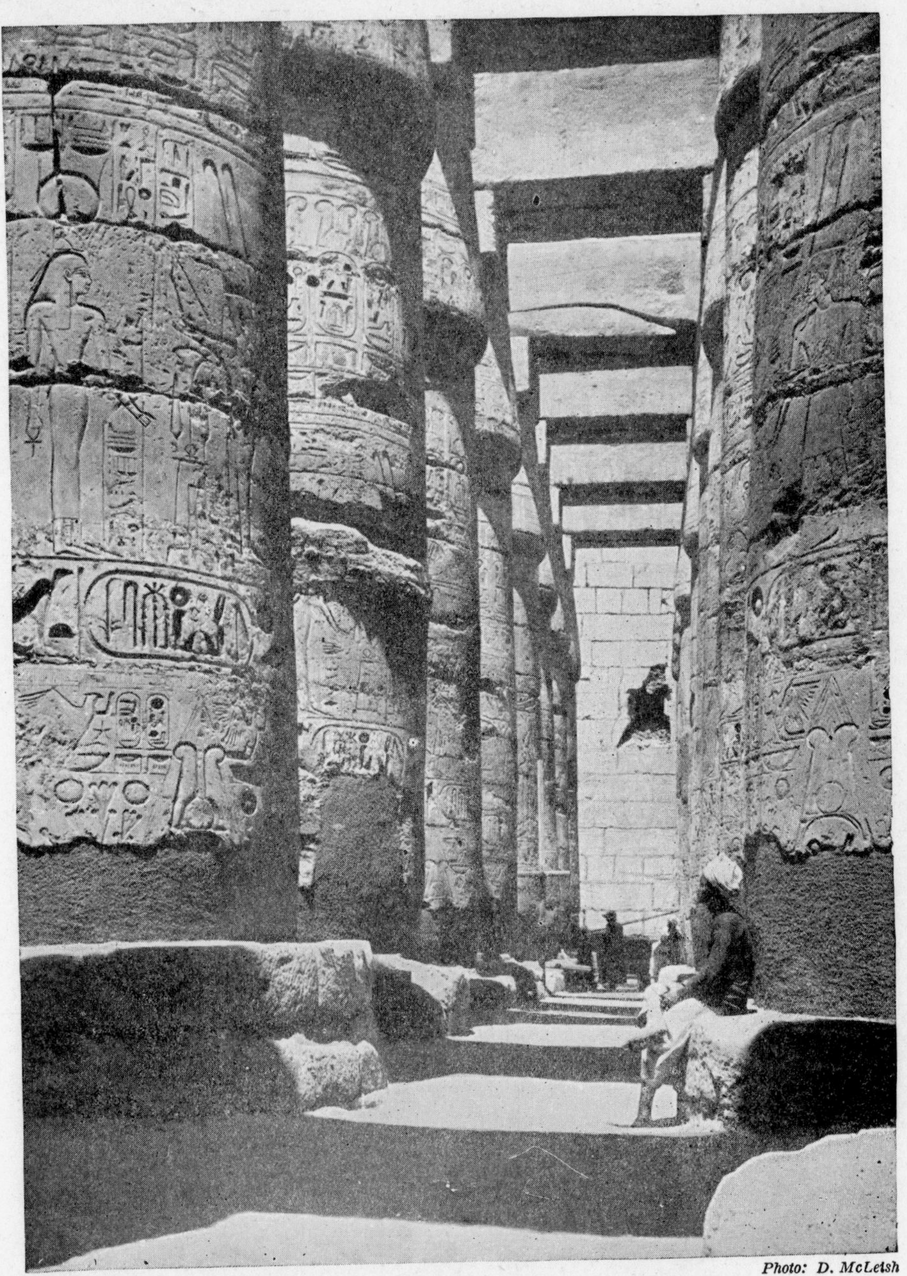

Photo: D. McLeish

THE GREAT HYPOSTYLE HALL AT KARNAK

the north-east and from the north-west. These would be the names of certain barbaric tribes, armed with iron weapons and using horse-chariots, who were becoming a great affliction to the Ægean and Semitic civilizations on the northern borders. They all spoke variants of what once must have been the same language, Aryan.

Round the north-east of the Black and Caspian Seas were coming the Medes and Persians. Confused with these in the records of the time were Scythians and Samatians. From north-east or north-west came the Armenians, from the north-west of the sea-barrier through the Balkan peninsula came Cimmerians, Phrygians and the Hellenic tribes whom now we call the Greeks. They were raiders and robbers and plunderers of cities, these Ayrans, east and west alike. They were all kindred and similar peoples, hardy herdsmen who had taken to plunder. In the east they were still only borderers and raiders, but in the west they were taking cities and driving out the civilized Ægean populations. The Ægean peoples were so pressed that they were seeking new homes in lands beyond the Aryan range. Some were seeking a settlement in the delta of the Nile and being repulsed by the Egyptians; some, the Etruscans, seem to have sailed from Asia Minor to found a state in the forest wildernesses of middle Italy; some built themselves cities upon the south-east coasts of the Mediterranean and became later that people known in history as the Philistines.

Of these Aryans who came thus rudely upon the scene of the ancient civilizations we will tell more fully in a later section. Here we note simply all this stir and emigration amidst the area of the ancient civilizations, that was set up by the swirl of the gradual and continuous advance of these Aryan barbarians out of the northern forests and wildernesses between 1600 and 600 B.C.

And in a section to follow we must tell also of a little Semitic people, the Hebrews, in the hills behind the Phœnician and Philistine coasts, who began to be of significance in the world towards the end of this period. They produced a literature of very great importance in subsequent history, a collection of books, histories, poems, books of wisdom and prophetic works, the Hebrew Bible.

In Mesopotamia and Egypt the coming of the Aryans did not cause fundamental changes until after 600 B.C. The flight of the

Ægeans before the Greeks and even the destruction of Cnossos must have seemed a very remote disturbance to both the citizens of Egypt and of Babylon. Dynasties came and went in these cradle states of civilization, but the main tenor of human life went on, with a slow increase in refinement and complexity age by age. In Egypt the accumulated monuments of more ancient times — the pyramids were already in their third thousand of years and a show for visitors just as they are to-day — were supplemented by fresh and splendid buildings, more particularly in the time of the seventeenth and nineteenth dynasties. The great temples at Karnak and Luxor date from this time. All the chief monuments of Nineveh, the great temples, the winged bulls with human heads, the reliefs of kings and

Photo: Jacques Boyer

FRIEZE SHOWING EGYPTIAN FEMALE SLAVES CARRYING LUXURIOUS FOODS

chariots and lion hunts, were done in these centuries between 1600 and 600 B.C., and this period also covers most of the splendours of Babylon.

Both from Mesopotamia and Egypt we now have abundant public records, business accounts, stories, poetry and private correspondence. We know that life, for prosperous and influential people in such cities as Babylon and the Egyptian Thebes, was already almost as refined and as luxurious as that of comfortable and prosperous people to-day. Such people lived an orderly and ceremonious life in beautiful and beautifully furnished and decorated houses, wore richly decorated clothing and lovely jewels; they had feasts and festivals, entertained one another with music and dancing, were waited upon by highly trained servants, were cared for by doctors and dentists. They did not travel very much or very far, but boating excursions were a common summer pleasure both on the Nile and on the Euphrates. The beast of burthen was the ass; the horse was still used only in chariots for war and upon occasions of state. The mule was still novel and the camel, though it was known in Mesopotamia, had not been brought into Egypt. And there were few utensils of iron; copper and bronze remained the prevailing metals. Fine linen and cotton fabrics were known as well as wool. But there was no silk yet. Glass was known and beautifully coloured, but glass things were usually small. There was no clear glass and no optical use of glass. People had gold stoppings in their teeth but no spectacles on their noses.

One odd contrast between the life of old Thebes or Babylon and modern life was the absence of coined money. Most trade was still done by barter. Babylon was financially far ahead of Egypt Gold and silver were used for exchange and kept in ingots; and there were bankers, before coinage, who stamped their names and the weight on these lumps of precious metal. A merchant or traveller would carry precious stones to sell to pay for his necessities. Most servants and workers were slaves who were paid not money but in kind. As money came in slavery declined.

A modern visitor to these crowning cities of the ancient world would have missed two very important articles of diet: there were no hens and no eggs. A French cook would have found small joy in Babylon. These things came from the East somewhere about the time of the last Assyrian empire.

Religion like everything else had undergone great refinement. Human sacrifice for instance had long since disappeared; animals or bread dummies had been substituted for the victim. (But the Phœnicians and especially the citizens of Carthage, their greatest settlement in Africa, were accused later of immolating human beings.) When a great chief had died in the ancient days it had been customary to sacrifice his wives and slaves and break spear and bow at his tomb so that he should not go unattended and unarmed in the spirit world. In Egypt there survived of this dark tradition the pleasant custom of burying small models of house and shop and servants and cattle with the dead, models that give us to-day the liveliest realization of the safe and cultivated life of these ancient people, three thousand years and more ago.

THE TEMPLE OF HORUS AT EDFU

Such was the ancient world before the coming of the Aryans out of the northern forests and plains. In India and China there were parallel developments. In the great valleys of both these regions agricultural city states of brownish peoples were growing up, but in India they do not seem to have advanced or coalesced so rapidly as the city states of Mesopotamia or Egypt. They were nearer the level of the ancient Sumerians or of the Maya civilization of America. Chinese history has still to be modernized by Chinese scholars and cleared of much legendary matter. Probably China at this time was in advance of India. Contemporary with the seventeenth dynasty in Egypt, there was a dynasty of emperors in China, the Shang dynasty, priest emperors over a loose-knit empire of subordinate kings. The chief duty of these early emperors was to perform the seasonal sacrifices. Beautiful bronze vessels from the time of the Shang dynasty still exist, and their beauty and workmanship compel us to recognize that many centuries of civilization must have preceded their manufacture.

XIX

The Primitive Aryans

FOUR thousand years ago, that is to say about 2000 B.C., central and south-eastern Europe and central Asia were probably warmer, moister and better wooded than they are now. In these regions of the earth wandered a group of tribes mainly of the fair and blue-eyed Nordic race, sufficiently in touch with one another to speak merely variations of one common language from the Rhine to the Caspian Sea. At that time they may not have been a very numerous people, and their existence was unsuspected by the Babylonians to whom Hammurabi was giving laws, or by the already ancient and cultivated land of Egypt which was tasting in those days for the first time the bitterness of foreign conquest.

These Nordic people were destined to play a very important part indeed in the world's history. They were a people of the parklands and the forest clearings; they had no horses at first but they had cattle; when they wandered they put their tents and other gear on rough ox waggons; when they settled for a time they may have made huts of wattle and mud. They burnt their important dead; they did not bury them ceremoniously as the brunette peoples did. They put the ashes of their greater leaders in urns and then made a great circular mound about them. These mounds are the "round barrows" that occur all over north Europe. The brunette people, their predecessors, did not burn their dead but buried them in a sitting position in elongated mounds; the "long barrows."

The Aryans raised crops of wheat, ploughing with oxen, but they did not settle down by their crops; they would reap and move on. They had bronze, and somewhen about 1500 B.C. they acquired iron. They may have been the discoverers of iron smelting. And somewhen vaguely about that time they also got the horse — which to begin with they used only for draught purposes. Their social life did not centre upon a temple like that of the more settled people round the Mediterranean, and their chief men were leaders rather than priests. They had an aristocratic social order rather than a

A BEAUTIFUL ARCHAIC AMPHORA

Compare the horses and other animals with the Altamira drawing on p. 54, and also with the Greek frieze, p. 140

divine and regal order; from a very early stage they distinguished certain families as leaderly and noble.

They were a very vocal people. They enlivened their wanderings by feasts, at which there was much drunkenness and at which a special sort of man, the bards, would sing and recite. They had no writing until they had come into contact with civilization, and the memories of these bards were their living literature. This use of recited language as an entertainment did much to make it a fine and beautiful instrument of expression, and to that no doubt the subsequent predominance of the languages derived from Aryan is, in part, to be ascribed. Every Aryan people had its legendary history crystallized in bardic recitations, epics, sagas and vedas, as they were variously called.

The social life of these people centred about the households of their leading men. The hall of the chief where they settled for a time was often a very capacious timber building. There were no doubt huts for herds and outlying farm buildings; but with most of the Aryan peoples this hall was the general centre, everyone went there to feast and hear the bards and take part in games and discussions. Cowsheds and stabling surrounded it. The chief and his wife and so forth would sleep on a dais or in an upper gallery; the commoner sort slept about anywhere, as people still do in Indian households. Except for weapons, ornaments, tools and suchlike personal possessions there was a sort of patriarchal communism in the tribe. The chief owned the cattle and grazing lands in the common interest; forest and rivers were the wild.

This was the fashion of the people who were increasing and multiplying over the great spaces of central Europe and west central Asia during the growth of the great civilization of Mesopotamia and the Nile, and whom we find pressing upon the heliolithic peoples everywhere in the second millennium before Christ. They were coming into France and Britain and into Spain. They pushed westward in two waves. The first of these people who reached Britain and Ireland were armed with bronze weapons. They exterminated or subjugated the people who had made the great stone monuments of Carnac in Brittany and Stonehenge and Avebury in England. They reached Ireland. They are called the Goidelic Celts. The second wave of a closely kindred people, perhaps intermixed with other racial elements, brought iron with it into Great Britain, and is known as the wave of Brythonic Celts. From them the Welsh derive their language.

Photo: Underwood & Underwood

THE MOUND OF NIPPUR

The site of a city which recent excavations have proved to date from at least as early as 5000 B.C., and probably 1000 years earlier

Kindred Celtic peoples were pressing southward into Spain and coming into contact not only with the heliolithic Basque people who still occupied the country but with the Semitic Phœnician colonies of the sea coast. A closely allied series of tribes, the Italians, were making their way down the still wild and wooded Italian peninsula. They did not always conquer. In the eighth century B.C. Rome appears in history, a trading town on the Tiber, inhabited by Aryan Latins but under the rule of Etruscan nobles and kings.

At the other extremity of the Aryan range there was a similar progress southward of similar tribes. Aryan peoples, speaking Sanskrit, had come down through the western passes into North

India long before 1000 B.C. There they came into contact with a primordial brunette civilization, the Dravidian civilization, and learnt much from it. Other Aryan tribes seem to have spread over the mountain masses of Central Asia far to the east of the present range of such peoples. In Eastern Turkestan there are still fair, blue-eyed Nordic tribes, but now they speak Mongolian tongues.

Between the Black and Caspian Seas the ancient Hittites had been submerged and "Aryanized" by the Armenians before 1000 B.C., and the Assyrians and Babylonians were already aware of a new and formidable fighting barbarism on the north-eastern frontiers, a group of tribes amidst which the Scythians, the Medes and the Persians remain as outstanding names.

But it was through the Balkan peninsula that Aryan tribes made their first heavy thrust into the heart of the old-world civilization. They were already coming southward and crossing into Asia Minor many centuries before 1000 B.C. First came a group of tribes of whom the Phrygians were the most conspicuous, and then in succession the Æolic, the Ionic and the Dorian Greeks. By 1000 B.C. they had wiped out the ancient Ægean civilization both in the mainland of Greece and in most of the Greek islands; the cities of Mycenæ and Tiryns were obliterated and Cnossos was nearly forgotten. The Greeks had taken to the sea before 1000 A.D., they had settled in Crete and Rhodes, and they were founding colonies in Sicily and the south of Italy after the fashion of the Phœnician trading cities that were dotted along the Mediterranean coasts.

So it was, while Tiglath Pileser III and Sargon II and Sardanapalus were ruling in Assyria and fighting with Babylonia and Syria and Egypt, the Aryan peoples were learning the methods of civilization and making it over for their own purposes in Italy and Greece and north Persia. The theme of history from the ninth century B.C. onward for six centuries is the story of how these Aryan peoples grew to power and enterprise and how at last they subjugated the whole Ancient World, Semitic, Ægean and Egyptian alike. In form the Aryan peoples were altogether victorious; but the struggle of Aryan, Semitic and Egyptian ideas and methods was continued long after the sceptre was in Aryan hands. It is indeed a struggle that goes on through all the rest of history and still in a manner continues to this day.

XX

The Last Babylonian Empire and the Empire of Darius I

We have already mentioned how Assyria became a great military power under Tiglath Pileser III and under the usurper Sargon II. Sargon was not this man's original name; he adopted it to flatter the conquered Babylonians by reminding them of that ancient founder of the Akkadian Empire, Sargon I, two thousand years before his time. Babylon, for all that it was a conquered city, was of greater population and importance than Nineveh, and its great god Bel Marduk and its traders and priests had to be treated politely. In Mesopotamia in the eighth century B.C. we are already far beyond the barbaric days when the capture of a town meant loot and massacre. Conquerors sought to propitiate and win the conquered. For a century and a half after Sargon the new Assyrian empire endured and, as we have noted, Assurbanipal (Sardanapalus) held at least lower Egypt.

But the power and solidarity of Assyria waned rapidly. Egypt by an effort threw off the foreigner under a Pharoah Psammetichus I, and under Necho II attempted a war of conquest in Syria. By that time Assyria was grappling with foes nearer at hand, and could make but a poor resistance. A Semitic people from south-east Mesopotamia, the Chaldeans, combined with Aryan Medes and Persians from the north-east against Nineveh, and in 606 B.C.—for now we are coming down to exact chronology—took that city.

There was a division of the spoils of Assyria. A Median Empire was set up in the north under Cyaxares. It included Nineveh, and its capital was Ecbatana. Eastward it reached to the borders of India. To the south of this in a great crescent was a new Chaldean Empire, the Second Babylonian Empire, which rose to a very great degree of wealth and power under the rule of Nebuchadnezzar the Great (the Nebuchadnezzar of the Bible). The last great days, the greatest days of all, for Babylon began. For a time the two Empires remained at peace, and the daughter of Nebuchadnezzar was married to Cyaxares.

Meanwhile Necho II was pursuing his easy conquests in Syria. He had defeated and slain King Josiah of Judah, a small country of which there is more to tell presently, at the battle of Megiddo in 608 B.C., and he pushed on to the Euphrates to encounter not a decadent Assyria but a renascent Babylonia. The Chaldeans dealt very vigorously with the Egyptians. Necho was routed and driven back to Egypt, and the Babylonian frontier pushed down to the ancient Egyptian boundaries.

From 606 until 539 B.C. the Second Babylonian Empire flourished insecurely. It flourished so long as it kept the peace with the stronger, hardier Median Empire to the north. And during these sixty-seven years not only life but learning flourished in the ancient city.

Even under the Assyrian monarchs and especially under Sardanapalus, Babylon had been a scene of great intellectual activity.

Map showing the relation of the MEDIAN and Second BABYLONIAN (Chaldæan) EMPIRES in the reign of Nebuchadnezzar the Great

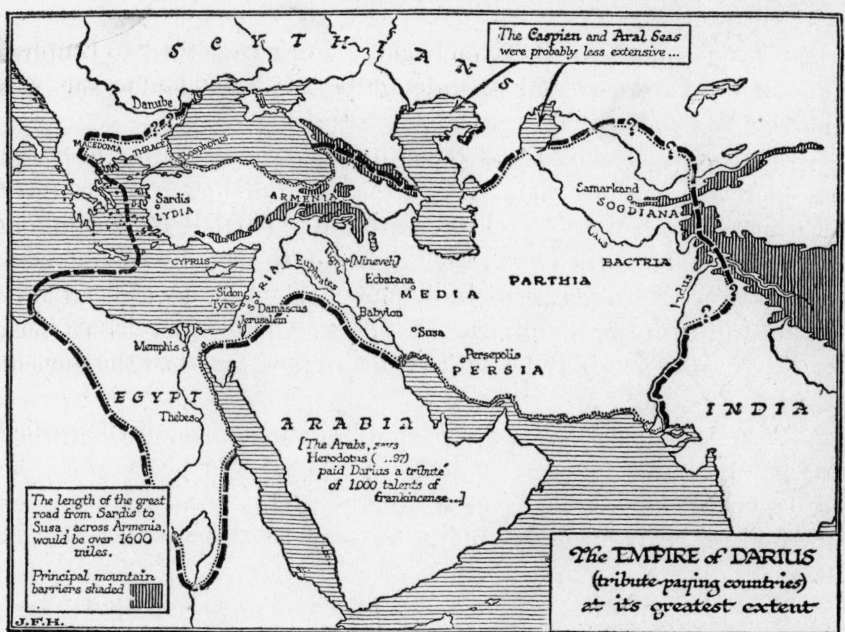

Sardanapalus, though an Assyrian, had been quite Babylon-ized. He made a library, a library not of paper but of the clay tablets that were used for writing in Mesopotamia since early Sumerian days. His collection has been unearthed and is perhaps the most precious store of historical material in the world. The last of the Chaldean line of Babylonian monarchs, Nabonidus, had even keener literary tastes. He patronized antiquarian researches, and when a date was worked out by his investigators for the accession of Sargon I he commemorated the fact by inscriptions. But there were many signs of disunion in his empire, and he sought to centralize it by bringing a number of the various local gods to Babylon and setting up temples to them there. This device was to be practised quite successfully by the Romans in later times, but in Babylon it roused the jealousy of the powerful priesthood of Bel Marduk, the dominant god of the Babylonians. They cast about for a possible alternative to Nabonidus and found it in Cyrus the Persian, the ruler of the adjacent Median Empire. Cyrus had already distinguished himself by conquering Crœsus, the rich king of Lydia in Eastern Asia Minor,

He came up against Babylon, there was a battle outside the walls, and the gates of the city were opened to him (538 B.C.). His soldiers entered the city without fighting. The crown prince Belshazzar, the son of Nabonidus, was feasting, the Bible relates, when a hand appeared and wrote in letters of fire upon the wall these mystical words: "*Mene, Mene, Tekel, Upharsin,*" which was interpreted by the prophet Daniel, whom he summoned to read the riddle, as "God has numbered thy kingdom and finished it; thou art weighed in the balance and found wanting and thy kingdom is given to the Medes and Persians." Possibly the priests of Bel Marduk knew something about that writing on the wall. Belshazzar was killed that night, says the Bible. Nabonidus was taken prisoner, and the occupation of the city was so peaceful that the services of Bel Marduk continued without intermission.

Photo: Miss J. Biggs

PERSIAN MONARCH

From the ruins of Persepolis

Thus it was the Babylonian and Median empires were united. Cambyses, the son of Cyrus, subjugated Egypt. Cambyses went mad and was accidentally killed, and was presently succeeded by Darius the Mede, Darius I, the son of Hystaspes, one of the chief councillors of Cyrus.

The Persian Empire of Darius I, the first of the new Aryan empires in the seat of the old civilizations, was the greatest empire the world had hitherto seen. It included all Asia Minor and Syria, all the old Assyrian and Babylonian empires, Egypt, the Caucasus and Caspian regions, Media, Persia, and it extended into India as far as the Indus. Such an empire was possible because the horse and rider and the chariot and the made-road had now been brought into the world. Hitherto the ass and ox and the camel for desert use had afforded the swiftest method of

Photo: Major W. J. P. Rodd

THE RUINS OF PERSEPOLIS

The capital city of the Persian Empire; burnt by Alexander the Great

Photo: Major W. J. P. Rodd

THE GREAT PORCH OF XERXES, AT PERSEPOLIS

transport. Great arterial roads were made by the Persian rulers to hold their new empire, and post horses were always in waiting for the imperial messenger or the traveller with an official permit. Moreover the world was now beginning to use coined money, which greatly facilitated trade and intercourse. But the capital of this vast empire was no longer Babylon. In the long run the priesthood of Bel Marduk gained nothing by their treason. Babylon though still important was now a declining city, and the great cities of the new empire were Persepolis and Susa and Ecbatana. The capital was Susa. Nineveh was already abandoned and sinking into ruins.

XXI

The Early History of the Jews

AND now we can tell of the Hebrews, a Semitic people, not so important in their own time as in their influence upon the later history of the world. They were settled in Judea long before 1000 B.C., and their capital city after that time was Jerusalem. Their story is interwoven with that of the great empires on either side of them, Egypt to the south and the changing empires of Syria, Assyria and Babylon to the north. Their country was an inevitable high road between these latter powers and Egypt.

Their importance in the world is due to the fact that they produced a written literature, a world history, a collection of laws, chronicles, psalms, books of wisdom, poetry and fiction and political utterances which became at last what Christians know as the Old Testament, the Hebrew Bible. This literature appears in history in the fourth or fifth century B.C.

Probably this literature was first put together in Babylon. We have already told how the Pharaoh, Necho II, invaded the Assyrian Empire while Assyria was fighting for life against Medes, Persians and Chaldeans. Josiah King of Judah opposed him, and was defeated and slain at Megiddo (608 B.C.). Judah became a tributary to Egypt, and when Nebuchadnezzar the Great, the new Chaldean king in Babylon, rolled back Necho into Egypt, he attempted to manage Judah by setting up puppet kings in Jerusalem. The experiment failed, the people massacred his Babylonian officials, and he then determined to break up this little state altogether, which had long been playing off Egypt against the northern empire. Jerusalem was sacked and burnt, and the remnant of the people was carried off captive to Babylon.

There they remained until Cyrus took Babylon (538 B.C.). He then collected them together and sent them back to resettle their country and rebuild the walls and temple of Jerusalem.

Before that time the Jews do not seem to have been a very civilized or united people. Probably only a very few of them could read or write. In their own history one never hears of the early books of the Bible being read; the first mention of a book is in the time of Josiah. The Babylonian captivity civilized them and consolidated them. They returned aware of their own literature, an acutely self-conscious and political people.

Their Bible at that time seems to have consisted only of the Pentateuch, that is to say the first five books of the Old Testament as we know it. In addition, as separate books they already had many of the other books that have since been incorporated with the Pentateuch into the present Hebrew Bible, Chronicles, the Psalms and Proverbs for example.

The accounts of the Creation of the World, of Adam and Eve and of the Flood, with which the Bible begins, run closely parallel with similar Babylonian legends; they seem to have been part of the common beliefs of all the Semitic peoples. So too the stories of Moses and of Samson have Sumerian and Babylonian parallels. But with the story of Abraham and onward begins something more special to the Jewish race.

Abraham may have lived as early as the days of Hammurabi in Babylon. He was a patriarchal Semitic nomad. To the book of Genesis the reader must go for the story of his wanderings and for the stories of his sons and grandchildren and how they became captive in the Land of Egypt. He travelled through Canaan, and the God of Abraham, says the Bible story, promised this smiling land of prosperous cities to him and to his children.

And after a long sojourn in Egypt and after fifty years of wandering in the wilderness under the leadership of Moses, the children of Abraham, grown now to a host of twelve tribes, invaded the land of Canaan from the Arabian deserts to the East. They may have done this somewhen between 1600 B.C. and 1300 B.C.; there are no Egyptian records of Moses nor of Canaan at this time to help out the story. But at any rate they did not succeed in conquering any more than the hilly backgrounds of the promised land. The coast was now in the hands, not of the Canaanites but of newcomers, those Ægean peoples, the Philistines; and their cities, Gaza, Gath, Ashdod, Ascalon and Joppa successfully withstood the Hebrew attack. For many generations the children of Abraham remained an obscure people of the hilly back country engaged in

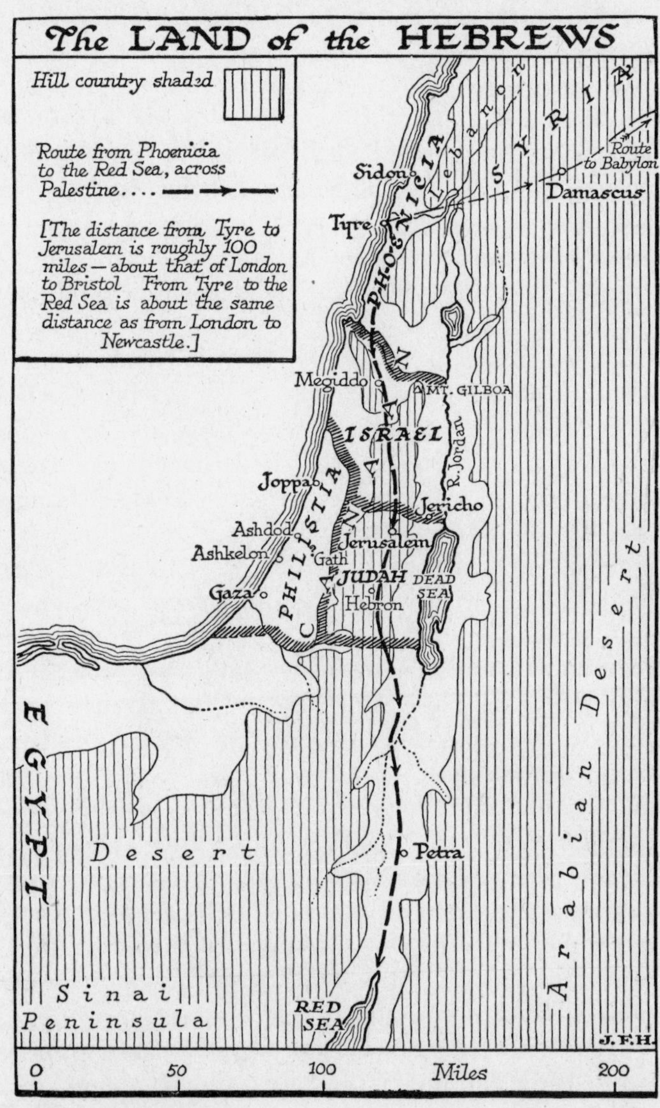

incessant bickerings with the Philistines and with the kindred tribes about them, the Moabites, the Midianites and so forth. The reader will find in the book of Judges a record of their struggles and disasters during this period. For very largely it is a record of disasters and failures frankly told.

For most of this period the Hebrews were ruled, so far as there was any rule among them, by priestly judges selected by the elders of the people, but at last somewhen towards 1000 B.C. they chose themselves a king, Saul, to lead them in battle. But Saul's leading was no great improvement upon the leading of the Judges; he perished under the hail of Philistine arrows at the battle of Mount Gilboa, his armour went into the temple of the Philistine Venus, and his body was nailed to the walls of Beth-shan.

His successor David was more successful and more politic. With David dawned the only period of prosperity the Hebrew peoples were ever to know. It was based on a close alliance with the Phœnician city of Tyre, whose King Hiram seems to have been a man of very great intelligence and enterprise. He wished to secure a trade route to the Red Sea through the Hebrew hill country. Normally Phœnician trade went to the Red Sea by Egypt, but Egypt was in a state of profound disorder at this

Photo: Underwood & Underwood

MOUND AT BABYLON

Beneath which are the remains of a great palace of Nebuchadnezzar time; there may have been other obstructions to Phœnician trade along this line, and at any rate Hiram established the very closest relations both with David and with his son and successor Solomon. Under Hiram's auspices the walls, palace and temple of Jerusalem arose, and in return Hiram built and launched his ships on the Red Sea. A very considerable trade passed northward and southward through Jerusalem. And Solomon achieved a prosperity and magnificence unprecedented in the experience of his people. He was even given a daughter of Pharaoh in marriage.

But it is well to keep the proportion of things in mind. At the climax of his glories Solomon was only a little subordinate king in a little city. His power was so transitory that within a few years of his death, Shishak the first Pharaoh of the twenty-second dynasty, had taken Jerusalem and looted most of its splendours. The account of Solomon's magnificence given in the books of Kings and Chronicles is questioned by many critics. They say that it was added to and exaggerated by the patriotic pride of later writers. But the Bible account read carefully is not so overwhelming as it appears at the first reading. Solomon's temple, if one works out the measurements, would go inside a small suburban church, and his fourteen hundred chariots cease to impress us when we learn from an Assyrian monument that his successor Ahab sent a contingent of two thousand to the Assyrian army. It is also plainly manifest from the Bible narrative that Solomon spent himself in display and overtaxed and overworked his people. At his death the northern part of his kingdom broke off from Jerusalem and became the independent kingdom of Israel. Jerusalem remained the capital city of Judah.

The prosperity of the Hebrew people was short-lived. Hiram died, and the help of Tyre ceased to strengthen Jerusalem. Egypt grew strong again. The history of the kings of Israel and the kings of Judah becomes a history of two little states ground between, first, Syria, then Assyria and then Babylon to the north and Egypt to the south. It is a tale of disasters and of deliverances that only delayed disaster. It is a tale of barbaric kings ruling a barbaric people. In 721 B.C. the kingdom of Israel was swept away into captivity by the Assyrians and its people utterly lost to history. Judah struggled

Photo: Underwood & Underwood

THE ISHTAR GATEWAY, BABYLON

The bulls are in richly coloured enamel on baked brick

on until in 604 B.C., as we have told, it shared the fate of Israel. There may be details open to criticism in the Bible story of Hebrew history from the days of the Judges onward, but on the whole it is evidently a true story which squares with all that has been learnt in the excavation of Egypt and Assyria and Babylon during the past century.

It was in Babylon that the Hebrew people got their history together and evolved their tradition. The people who came back to Jerusalem at the command of Cyrus were a very different people in spirit and knowledge from those who had gone into captivity. They had learnt civilization. In the development of their peculiar character a very great part was played by certain men, a new sort of men, the Prophets, to whom we must now direct our attention. These Prophets mark the appearance of new and remarkable forces in the steady development of human society.

XXII

Priests and Prophets in Judea

THE fall of Assyria and Babylon were only the first of a series of disasters that were to happen to the Semitic peoples. In the seventh century B.C. it would have seemed as though the whole civilized world was to be dominated by Semitic rulers. They ruled the great Assyrian empire and they had conquered Egypt; Assyria, Babylon, Syria were all Semitic, speaking languages that were mutually intelligible. The trade of the world was in Semitic hands. Tyre, Sidon, the great mother cities of the Phœnician coast, had thrown out colonies that grew at last to even greater proportion in Spain, Sicily and Africa. Carthage, founded before 800 B.C., had risen to a population of more than a million. It was for a time the greatest city on earth. Its ships went to Britain and out into the Atlantic. They may have reached Madeira. We have already noted how Hiram co-operated with Solomon to build ships on the Red Sea for the Arabian and perhaps for the Indian trade. In the time of the Pharaoh Necho, a Phœnician expedition sailed completely round Africa.

At that time the Aryan peoples were still barbarians. Only the Greeks were reconstructing a new civilization of the ruins of the one they had destroyed, and the Medes were becoming "formidable," as an Assyrian inscription calls them, in central Asia. In 800 B.C. no one could have prophesied that before the third century B.C. every trace of Semitic dominion would be wiped out by Aryan-speaking conquerors, and that everywhere the Semitic peoples would be subjects or tributaries or scattered altogether. Everywhere except in the northern deserts of Arabia, where the Bedouin adhered steadily to the nomadic way of life, the ancient way of life of the Semites before Sargon I and his Akkadians went down to conquer Sumeria. But the Arab Bedouin were never conquered by Aryan masters.

Now of all these civilized Semites who were beaten and overrun in these five eventful centuries one people only held together and clung to its ancient traditions and that was this little people, the Jews, who were sent back to build their city of Jerusalem by Cyrus the Persian. And they were able to do this, because they had got together this literature of theirs, their Bible, in Babylon. It is not so much the Jews who made the Bible as the Bible which made the Jews. Running through this Bible were certain ideas, different from the ideas of the people about them, very stimulating and sustaining ideas, to which they were destined to cling through five and twenty centuries of hardship, adventure and oppression.

Foremost of these Jewish ideas was this, that their God was invisible and remote, an invisible God in a temple not made with hands, a Lord of Righteousness throughout the earth. All other peoples had national gods embodied in images that lived in temples. If the image was smashed and the temple razed, presently that god died out. But this was a new idea, this God of the Jews, in the heavens, high above priests and sacrifices. And this God of Abraham, the Jews believed, had chosen them to be his peculiar people, to restore Jerusalem and make it the capital of Righteousness in the World. They were a people exalted by their sense of a common destiny. This belief saturated them all when they returned to Jerusalem after the captivity in Babylon.

Is it any miracle that in their days of overthrow and subjugation many Babylonians and Syrians and so forth and later on many Phœnicians, speaking practically the same language and having endless customs, habits, tastes and traditions in common, should be attracted by this inspiring cult and should seek to share in its fellowship and its promise? After the fall of Tyre, Sidon, Carthage and the Spanish Phœnician cities, the Phœnicians suddenly vanish from history; and as suddenly we find, not simply in Jerusalem but in Spain, Africa, Egypt, Arabia, the East, wherever the Phœnicians had set their feet, communities of Jews. And they were all held together by the Bible and by the reading of the Bible. Jerusalem was from the first only their nominal capital; their real city was this book of books. This is a new sort of thing in history. It is something of which the seeds were sown long before, when the Sumerians and Egyptians began to turn their hieroglyphics into writing. The Jews were a new thing, a people without a king and presently without a temple (for as we shall tell Jerusalem itself was broken up in 70 A.D.), held together and consolidated out of heterogeneous elements by nothing but the power of the written word.

And this mental welding of the Jews was neither planned nor foreseen nor done by either priests or statesmen. Not only a new kind of community but a new kind of man comes into history with the development of the Jews. In the days of Solomon the Hebrews looked like becoming a little people just like any other little people of that time clustering around court and temple, ruled by the wisdom of the priest and led by the ambition of the king. But already, the reader may learn from the Bible, this new sort of man of which we speak, the Prophet, was in evidence.

As troubles thicken round the divided Hebrews the importance of these Prophets increases.

What were these Prophets? They were men of the most diverse origins. The Prophet Ezekiel was of the priestly caste and the Prophet Amos wore the goatskin mantle of a shepherd, but all had this in common, that they gave allegiance to no one but to the God of Righteousness and that they spoke directly to the people. They

THE BLACK OBELISK OF SHALMANESER II

This obelisk (in the British Museum) of the King of Assyria mentions, in cuneiform, "Jehu the son of Omri." Panel showing Jewish captives bringing tribute

ANOTHER PANEL OF THE BLACK OBELISK
Captive Princes making obeisance to Shalmaneser II

came without licence or consecration. "Now the word of the Lord came unto me;" that was the formula. They were intensely political. They exhorted the people against Egypt, "that broken reed," or against Assyria or Babylon; they denounced the indolence of the priestly order or the flagrant sins of the King. Some of them turned their attention to what we should now call "social reform." The rich were "grinding the faces of the poor," the luxurious were consuming the children's bread; wealthy people made friends with and imitated the splendours and vices of foreigners; and this was hateful to Jehovah, the God of Abraham, who would certainly punish this land.

These fulminations were written down and preserved and studied. They went wherever the Jews went, and wherever they went they spread a new religious spirit. They carried the common man past priest and temple, past court and king and brought him face to face with the Rule of Righteousness. That is their supreme importance in the history of mankind. In the great utterances of Isaiah the prophetic voice rises to a pitch of splendid anticipation and foreshadows the whole earth united and at peace under one God. Therein the Jewish prophecies culminate.

All the Prophets did not speak in this fashion, and the intelligent reader of the prophetic books will find much hate in them, much prejudice, and much that will remind him of the propaganda pamphlets of the present time. Nevertheless it is the Hebrew Prophets of the period round and about the Babylonian captivity who mark the appearance of a new power in the world, the power of individual moral appeal, of an appeal to the free conscience of mankind against the fetish sacrifices and slavish loyalties that had hitherto bridled and harnessed our race.

XXIII

The Greeks

NOW while after Solomon (whose reign was probably about 960 B.C.) the divided kingdoms of Israel and Judah were suffering destruction and deportation, and while the Jewish people were developing their tradition in captivity in Babylon, another great power over the human mind, the Greek tradition, was also arising. While the Hebrew prophets were working out a new sense of direct moral responsibility between the people and an eternal and universal God of Right, the Greek philosophers were training the human mind in a new method and spirit of intellectual adventure.

The Greek tribes as we have told were a branch of the Aryan-speaking stem. They had come down among the Ægean cities and islands some centuries before 1000 B.C. They were probably already in southward movement before the Pharaoh Thothmes hunted his first elephants beyond the conquered Euphrates. For in those days there were elephants in Mesopotamia and lions in Greece.

It is possible that it was a Greek raid that burnt Cnossos, but there are no Greek legends of such a victory though there are stories of Minos and his palace (the Labyrinth) and of the skill of the Cretan artificers.

Like most of the Aryans these Greeks had singers and reciters whose performances were an important social link, and these handed down from the barbaric beginnings of their people two great epics, the *Iliad*, telling how a league of Greek tribes besieged and took and sacked the town of Troy in Asia Minor, and the *Odyssey*, being a long adventure story of the return of the sage captain, Odysseus, from Troy to his own island. These epics were written down somewhen in the eighth or seventh century B.C., when the Greeks had acquired the use of an alphabet from their more civilized neighbours, but they

Photo: Sebah & Jonillier

STATUE OF MELEAGER

Note the progress in plastic power from the earlier wooden statue on left

are supposed to have been in existence very much earlier. Formerly they were ascribed to a particular blind bard, Homer, who was supposed to have sat down and composed them as Milton composed *Paradise Lost.* Whether there really was such a poet, whether he composed or only wrote down and polished these epics and so forth, is a favourite quarrelling ground for the erudite. We need not concern ourselves with such bickerings here. The thing that matters from our point of view is that the Greeks were in possession of their epics in the eighth century B.C., and that they were a common possession and a link between their various tribes, giving them a sense of fellowship as against the outer barbarians. They were a group of kindred peoples linked by the spoken and afterwards by the written word, and sharing common ideals of courage and behaviour.

The epics showed the Greeks a barbaric people without iron, without writing, and still not living in cities. They seem to have lived at first in open villages of huts around the halls of their chiefs outside the ruins of the Ægean cities they had destroyed. Then they began to wall their cities and to adopt the idea of temples from the people they had conquered. It has been said that the cities of the primitive civilizations grew up about the altar of some tribal god, and that the wall was added; in the cities of the Greeks the wall preceded the temple. They began to trade and send out colonies. By the seventh century B.C. a new series of cities had grown up in the valleys and islands of Greece, forgetful of the Ægean cities and civilization that had preceded them; Athens, Sparta, Corinth, Thebes, Samos, Miletus among the chief. There were already Greek settlements along the coast of the Black Sea and in Italy and Sicily. The heel and toe of Italy was called Magna Græcia. Marseilles was a Greek town established on the site of an earlier Phœnician colony.

Now countries which are great plains or which have as a chief means of transport some great river like the Euphrates or Nile tend to become united under some common rule. The cities of Egypt and the cities of Sumeria, for example, ran together under one system of government. But the Greek peoples were cut up among islands and mountain valleys; both Greece and Magna Græcia are very mountainous; and the tendency was all the other way. When the

Greeks come into history they are divided up into a number of little states which showed no signs of coalescence. They are different even in race. Some consist chiefly of citizens of this or that Greek tribe, Ionic, Æolian or Doric; some have a mingled population of Greeks and descendants of the pre-Greek "Mediterranean" folk; some have an unmixed free citizenship of Greeks lording it over an enslaved conquered population like the "Helots" in Sparta. In some the old leaderly Aryan families have become a close aristocracy;

Photo: Fred Boissonnas

RUINS OF THE GREAT TEMPLE OF ZEUS AT OLYMPIA

in some there is a democracy of all the Aryan citizens; in some there are elected or even hereditary kings, in some usurpers or tyrants.

And the same geographical conditions that kept the Greek states divided and various, kept them small. The largest states were smaller than many English counties, and it is doubtful if the population of any of their cities ever exceeded a third of a million. Few came up even to 50,000. There were unions of interest and sympathy but no coalescences. Cities made leagues and alliances as trade increased, and small cities put themselves under the protection of great ones. Yet all Greece was held together in a certain community of feeling by two things, by the epics and by the custom of taking part every fourth year in the athletic contests at Olympia. This did not prevent wars and feuds, but it mitigated something of the savagery of war between them, and a truce protected all travellers to and from the games. As time went on the sentiment of a common heritage grew and the number of states participating in the Olympic games increased until at last not only Greeks but competitors from the closely kindred countries of Epirus and Macedonia to the north were admitted.

The Greek cities grew in trade and importance, and the quality of their civilization rose steadily in the seventh and sixth centuries B.C. Their social life differed in many interesting points from the social life of the Ægean and river valley civilizations. They had splendid temples but the priesthood was not the great traditional body it was in the cities of the older world, the repository of all knowledge, the storehouse of ideas. They had leaders and noble families, but no quasi-divine monarch surrounded by an elaborately organized court. Rather their organization was aristocratic, with leading families which kept each other in order. Even their so-called "democracies" were aristocratic; every citizen had a share in public affairs and came to the assembly in a democracy, *but everybody was not a citizen.* The Greek democracies were not like our modern "democracies" in which everyone has a vote. Many of the Greek democracies had a few hundred or a few thousand citizens and then many thousands of slaves, freedmen and so forth, with no share in public affairs. Generally in Greece affairs were in the hands of a community of substantial men. Their kings and their tyrants alike were just men set in front of other men or usurping a leadership; they were not quasi-divine overmen like Pharaoh or Minos or the monarchs of Mesopotamia. Both thought and government therefore had a freedom under Greek conditions such as they had known in none of the older civilizations. The Greeks had brought down into cities the individualism, the personal initiative of the wandering life of the northern parklands. They were the first republicans of importance in history.

And we find that as they emerge from a condition of barbaric warfare a new thing becomes apparent in their intellectual life. We find men who are not priests seeking and recording knowledge and enquiring into the mysteries of life and being, in a way that has hitherto been the sublime privilege of priesthood or the presumptu-

Photo: Alinari

THE TEMPLE OF NEPTUNE (POSEIDON), PÆSTUM, SICILY

ous amusement of kings. We find already in the sixth century B.C. — perhaps while Isaiah was still prophesying in Babylon — such men as Thales and Anaximander of Miletus and Heraclitus of Ephesus, who were what we should now call independent gentlemen, giving their minds to shrewd questionings of the world in which we live, asking what its real nature was, whence it came and what its destiny might be, and refusing all ready-made or evasive answers. Of these questionings of the universe by the Greek mind, we shall have more to say a little later in this history. These Greek enquirers who begin to be remarkable in the sixth century B.C. are the first philosophers, the first "wisdom-lovers," in the world.

And it may be noted here how important a century this sixth century B.C. was in the history of humanity. For not only were these Greek philosophers beginning the research for clear ideas about this universe and man's place in it and Isaiah carrying Jewish prophecy to its sublimest levels, but as we shall tell later Gautama Buddha was then teaching in India and Confucius and Lao Tse in China. From Athens to the Pacific the human mind was astir.

XXIV

The Wars of the Greeks and Persians

WHILE the Greeks in the cities in Greece, South Italy and Asia Minor were embarking upon free intellectual enquiry and while in Babylon and Jerusalem the last of the Hebrew prophets were creating a free conscience for mankind, two adventurous Aryan peoples, the Medes and the Persians, were in possession of the civilization of the ancient world and were making a great empire, the Persian empire, which was far larger in extent than any empire the world had seen hitherto. Under Cyrus, Babylon and the rich and ancient civilization of Lydia had been added to the Persian rule; the Phœnician cities of the Levant and all the Greek cities in Asia Minor had been made tributary, Cambyses had subjected Egypt, and Darius I, the Mede, the third of the Persian rulers (521 B.C.), found himself monarch as it seemed of all the world. His couriers rode with his decrees from the Dardanelles to the Indus and from Upper Egypt to Central Asia.

The Greeks in Europe, it is true, Italy, Carthage, Sicily and the Spanish Phœnician settlements, were not under the Persian Peace; but they treated it with respect and the only people who gave any serious trouble were the old parent hordes of Nordic people in South Russia and Central Asia, the Scythians, who raided the northern and north-eastern borders.

Of course the population of this great Persian empire was not a population of Persians. The Persians were only the small conquering minority of this enormous realm. The rest of the population was what it had been before the Persians came from time immemorial, only that Persian was the administrative language. Trade and finance were still largely Semitic, Tyre and Sidon as of old were the great Mediterranean ports and Semitic shipping plied upon the seas. But many of these Semitic merchants and business people as

Brit. Mus.

FINE PIECE OF ATHENIAN POTTERY

Showing Greek merchant vessels with sails and oars

they went from place to place already found a sympathetic and convenient common history in the Hebrew tradition and the Hebrew scriptures. A new element which was increasing rapidly in this empire was the Greek element. The Greeks were becoming serious rivals to the Semites upon the sea, and their detached and vigorous intelligence made them useful and unprejudiced officials.

It was on account of the Scythians that Darius I invaded Europe. He wanted to reach South Russia, the homeland of the Scythian horsemen. He crossed the Bosphorus with a great army and marched through Bulgaria to the Danube, crossed this by a bridge of boats and pushed far northward. His army suffered terribly. It was largely an infantry force and the mounted Scythians rode all round it, cut off its supplies, destroyed any stragglers and never came to a pitched battle. Darius was forced into an inglorious retreat.

He returned himself to Susa but he left an army in Thrace and Macedonia, and Macedonia submitted to Darius. Insurrections of the Greek cities in Asia followed this failure, and the European Greeks were drawn into the contest. Darius resolved upon the subjugation of the Greeks in Europe. With the Phœnician fleet at his disposal he was able to subdue one island after another, and finally in 490 B.C. he made his main attack upon Athens. A considerable Armada sailed from the ports of Asia Minor and the eastern Mediterranean, and the expedition landed its troops at Marathon to the north of Athens. There they were met and signally defeated by the Athenians.

An extraordinary thing happened at this time. The bitterest rival of Athens in Greece was Sparta, but now Athens appealed to Sparta, sending a herald, a swift runner, imploring the Spartans not to let Greeks become slaves to barbarians. This runner (the prototype of all "Marathon" runners) did over a hundred miles of broken country in less than two days. The Spartans responded promptly and generously; but when, in three days, the Spartan force reached Athens, there was nothing for it to do but to view the battlefield and the bodies of the defeated Persian soldiers. The Persian fleet had returned to Asia. So ended the first Persian attack on Greece.

The next was much more impressive. Darius died soon after the news of his defeat at Marathon reached him, and for four years his son and successor, Xerxes, prepared a host to crush the Greeks. For a time terror united all the Greeks. The army of Xerxes was certainly the greatest that had hitherto been assembled in the world. It was a huge assembly of discordant elements. It crossed the Dardanelles, 480 B.C., by a bridge of boats; and along the coast as it advanced moved an equally miscellaneous fleet carrying supplies. At the narrow pass of Thermopylæ a small force of 1400 men under the Spartan Leonidas resisted this multitude, and after a fight of unsurpassed heroism was completely destroyed. Every man was killed. But the losses they inflicted upon the Persians were enormous, and the army of Xerxes pushed on to Thebes and Athens in a chastened mood. Thebes surrendered and made terms. The Athenians abandoned their city and it was burnt.

Greece seemed in the hands of the conqueror, but again came victory against the odds and all expectations. The Greek fleet, though not a third the size of the Persian, assailed it in the bay of Salamis and destroyed it. Xerxes found himself and his immense army cut off from supplies and his heart failed him. He retreated to Asia with one half of his army, leaving the rest to be defeated at Platea (479 B.C.) what time the remnants of the Persian fleet were hunted down by the Greeks and destroyed at Mycalæ in Asia Minor.

The Persian danger was at an end. Most of the Greek cities in Asia became free. All this is told in great detail and with much picturesqueness in the first of written histories, the *History* of

Photo: Fred Boissonnas

ALL THAT REMAINS OF THE GREAT TEMPLE OF CORINTH

Photo: Fred Boissonnas

THE TEMPLE OF NEPTUNE (POSEIDON) AT CAPE SUNIUM

Herodotus. This Herodotus was born about 484 B.C. in the Ionian city of Halicarnassus in Asia Minor, and he visited Babylon and Egypt in his search for exact particulars. From Mycalæ onward Persia sank into a confusion of dynastic troubles. Xerxes was murdered in 465 B.C. and rebellions in Egypt, Syria and Media broke up the brief order of that mighty realm. The history of Herodotus lays stress on the weakness of Persia. This history is indeed what we should now call propaganda — propaganda for Greece to unite and conquer Persia. Herodotus makes one character, Aristagoras, go to the Spartans with a map of the known world and say to them: "These Barbarians are not valiant in fight. You on the other hand have now attained the utmost skill in war. . . . No other nations in the world have what they possess: gold, silver, bronze, embroidered garments, beasts and slaves. *All this you might have for yourselves, if you so desired.*"

XXV

The Splendour of Greece

THE century and a half that followed the defeat of Persia was one of very great splendour for the Greek civilization. True that Greece was torn by a desperate struggle for ascendancy between Athens, Sparta and other states (the Peloponnesian War 431 to 404 B.C.) and that in 338 B.C. the Macedonians became virtually masters of Greece; nevertheless during this period the thought and the creative and artistic impulse of the Greeks rose to levels that made their achievement a lamp to mankind for all the rest of history.

The head and centre of this mental activity was Athens. For over thirty years (466 to 428 B.C.) Athens was dominated by a man of great vigour and liberality of mind, Pericles, who set himself to rebuild the city from the ashes to which the Persians had reduced it. The beautiful ruins that still glorify Athens to-day are chiefly the remains of this great effort. And he did not simply rebuild a material Athens. He rebuilt Athens intellectually. He gathered about him not only architects and sculptors but poets, dramatists, philosophers and teachers. Herodotus came to Athens to recite his history (438 B.C.). Anaxagoras came with the beginnings of a scientific description of the sun and stars. Æschylus, Sophocles and Euripides one after the other carried the Greek drama to its highest levels of beauty and nobility.

The impetus Pericles gave to the intellectual life of Athens lived on after his death, and in spite of the fact that the peace of Greece was now broken by the Peloponnesian War and a long and wasteful struggle for "ascendancy" was beginning. Indeed the darkling of the political horizon seems for a time to have quickened rather than discouraged men's minds.

Already long before the time of Pericles the peculiar freedom of Greek institutions had given great importance to skill in discussion.

Decision rested neither with king nor with priest but in the assemblies of the people or of leading men. Eloquence and able argument became very desirable accomplishments therefore, and a class of teachers arose, the Sophists, who undertook to strengthen young men in these arts. But one cannot reason without matter, and knowledge followed in the wake of speech. The activities and rivalries of these Sophists led very naturally to an acute examination of style, of methods of thought and of the validity of arguments. When Pericles died a certain Socrates was becoming prominent as an able and destructive critic of bad argument—and much of the teaching of the Sophists was bad argument. A group of brilliant young men gathered about Socrates. In the end Socrates was executed for disturbing people's minds (399 B.C.), he was condemned after the dignified fashion of the Athens of those days to drink in his own house and among his own friends a poisonous draught made from hemlock, but the disturbance of people's minds went on in spite of his condemnation. His young men carried on his teaching.

Photo: Fred Boissonnas

PART OF THE FAMOUS FRIEZE OF THE PARTHENON, ATHENS

A specimen of Grecian sculpture in its finest expression. Compare the advance of art with that seen in the animals shown on p. 105

Chief among these young men was Plato (427 to 347 B.C.) who presently began to teach philosophy in the grove of the Academy. His teaching fell into two main divisions, an examination of the foundations and methods of human thinking and an examination of political institutions. He was the first man to write a Utopia, that is to say the plan of a community different from and better than any

Photo: Fred Boissonnas

THE ACROPOLIS, ATHENS

The marvellous group of Temples and monuments built under the inspiration of Pericles

Photo: Fred Boissonnas

THE THEATRE AT EPIDAUROS, GREECE

A wonderfully preserved specimen showing the vast auditorium

existing community. This shows an altogether unprecedented boldness in the human mind which had hitherto accepted social traditions and usages with scarcely a question. Plato said plainly to mankind: "Most of the social and political ills from which you suffer are under your control, given only the will and courage to change them. You can live in another and a wiser fashion if you choose to think it out and work it out. You are not awake to your own power." That is a high adventurous teaching that has still to soak in to the common intell'gence of our race. One of his earliest works was the *Republic*, a dream of a communist aristocracy; his last unfinished work was the *Laws*, a scheme of regulation for another such Utopian state.

The criticism of methods of thinking and methods of government was carried on after Plato's death by Aristotle, who had been his pupil and who taught in the Lyceum. Aristotle came from the city of Stagira in Macedonia, and his father was court physician to the Macedonian king. For a time Aristotle was tutor to Alex-

Photo: Fred Boissonnas

THE CARYATIDES OF THE ERECHTHEUM

The ancient sanctuary on the Acropolis at Athens

Photo: Alinari

ATHENE OF THE PARTHENON

ander, the king's son, who was destined to achieve very great things of which we shall soon be telling. Aristotle's work upon methods of thinking carried the science of Logic to a level at which it remained for fifteen hundred years or more, until the mediæval schoolmen took up the ancient questions again. He made no Utopias. Before man could really control his destiny as Plato taught, Aristotle perceived that he needed far more knowledge and far more accurate knowledge than he possessed. And so Aristotle began that systematic collection of knowledge which nowadays we call Science. He sent out explorers to collect *facts*. He was the father of natural history. He was the founder of political science. His students at the Lyceum examined and compared the constitutions of 158 different states. . . .

Here in the fourth century B.C. we find men who are practically "modern thinkers." The child-like, dream-like methods of primitive thought had given way to a disciplined and critical attack upon the problems of life. The weird and monstrous symbolism and imagery of the gods and god monsters, and all the taboos and awes and restraints that have hitherto encumbered thinking are here completely set aside. Free, exact and systematic thinking has begun. The fresh and unencumbered mind of these newcomers out of the northern forests has thrust itself into the mysteries of the temple and let the daylight in.

XXVI

The Empire of Alexander the Great

FROM 431 to 404 B.C. the Peloponnesian War wasted Greece. Meanwhile to the north of Greece, the kindred country of Macedonia was rising slowly to power and civilization. The Macedonians spoke a language closely akin to Greek, and on several occasions Macedonian competitors had taken part in the Olympic games. In 359 B.C. a man of very great abilities and ambition became king of this litt e country — Philip. Philip had previously been a hostage in Greece; he had had a thoroughly Greek education and he was probably aware of the ideas of Herodotus — which had also been developed by the philosopher Isocrates — of a possible conquest of Asia by a consolidated Greece.

He set himself first to extend and organize his own realm and to remodel his army. For a thousand years now the charging horse-chariot had been the decisive factor in battles, that and the close-fighting infantry. Mounted horsemen had also fought, but as a cloud of skirmishers, individually and without discipline. Philip made his infantry fight in a closely packed mass, the Macedonian phalanx, and he trained his mounted gentlemen, the knights or companions, to fight in formation and so invented cavalry. The master move in most of his battles and in the battles of his son Alexander was a cavalry charge. The phalanx *held* the enemy infantry in front while the cavalry swept away the enemy horse on his wings and poured in on the flank and rear of his infantry. Chariots were disabled by bowmen, who shot the horses.

With this new army Philip extended his frontiers through Thessaly to Greece; and the battle of Chæronia (338 B.C.), fought against Athens and her allies, put all Greece at his feet. At last the dream of Herodotus was bearing fruit. A congress of all the Greek states appointed Philip captain-general of the Græco-Macedonian confederacy against Persia, and in 336 B.C. his advanced guard crossed into Asia upon this long premeditated adventure. But he never followed it. He was assassinated; it is believed at the instigation of his queen Olympias, Alexander's mother. She was jealous because Philip had married a second wife.

BUST OF ALEXANDER THE GREAT
(As in the British Museum)

But Philip had taken unusual pains with his son's education. He had not only secured Aristotle, the greatest philosopher in the world, as this boy's tutor, but he had shared his ideas with him and thrust military experience upon him. At Chæronia Alexander, who was then only eighteen years old, had been in command of the cavalry. And so it was possible for this young man, who was still only twenty years old at the time of his accession, to take up his father's task at once and to proceed successfully with the Persian adventure.

In 334 B.C.—for two years were needed to establish and confirm his position in Macedonia and Greece—he crossed into Asia, defeated a not very much bigger Persian army at the battle of the Granicus and captured a number of cities in Asia Minor. He kept along the sea-coast. It was necessary for him to reduce and garrison all the coast towns as he advanced because the Persians had control of the fleets of Tyre and Sidon and so had command of the sea.

Had he left a hostile port in his rear the Persians might have landed forces to raid his communications and cut him off. At Issus (333 B.C.) he met and smashed a vast conglomerate host under Darius III. Like the host of Xerxes that had crossed the Dardanelles a century and a half before, it was an incoherent accumulation of contingents and it was encumbered with a multitude of court officials, the harem of Darius and many camp followers. Sidon surrendered to Alexander but Tyre resisted obstinately. Finally that great city was stormed and plundered and destroyed. Gaza

ALEXANDER'S VICTORY OVER THE PERSIANS AT ISSUS
(*From the Pompeian Mosaic*)
Alexander charges in on the left, Darius is in the chariot to the right

also was stormed, and towards the end of 332 B.C. the conqueror entered Egypt and took over its rule from the Persians.

At Alexandretta and at Alexandria in Egypt he built great cities, accessible from the land and so incapable of revolt. To these the trade of the Phœnician cities was diverted. The Phœnicians of the western Mediterranean suddenly disappear from history — and as immediately the Jews of Alexandria and the other new trading cities created by Alexander appear.

In 331 B.C. Alexander marched out of Egypt upon Babylon as Thothmes and Rameses and Necho had done before him. But he marched by way of Tyre. At Arbela near the ruins of Nineveh,

Photo: Alinari

THE APOLLO BELVEDERE

(In the Vatican Museum)

which was already a forgotten city, he met Darius and fought the decisive battle of the war. The Persian chariot charge failed, a Macedonian cavalry charge broke up the great composite host and the phalanx completed the victory. Darius led the retreat. He made no further attempt to resist the invader but fled northward into the country of the Medes. Alexander marched on to Babylon, still prosperous and important, and then to Susa and Persepolis. There after a drunken festival he burnt down the palace of Darius, the king of kings.

Thence Alexander presently made a military parade of central Asia, going to the utmost bounds of the Persian empire. At first he turned northward. Darius was pursued; and he was overtaken at dawn dying in his chariot, having been murdered by his own people. He was still living when the foremost Greeks reached him. Alexander came up to find him dead. Alexander skirted the Caspian Sea, he went up into the mountains of western Turkestan, he came down by Herat (which he founded) and Cabul and the Khyber Pass into

India. He fought a great battle on the Indus with an Indian king, Porus, and here the Macedonian troops met elephants for the first time and defeated them. Finally he built himself ships, sailed down to the mouth of the Indus, and marched back by the coast of Beluchistan, reaching Susa again in 324 B.C. after an absence of six years. He then prepared to consolidate and organize this vast empire he had won. He sought to win over his new subjects. He assumed the robes and tiara of a Persian monarch, and this roused the jealousy of his Macedonian commanders. He had much trouble with them. He arranged a number of marriages between these Macedonian officers and Persian and Babylonian women: the "Marriage of the East and West." He never lived to effect the consolidation he had planned. A fever seized him after a drinking bout in Babylon and he died in 323 B.C.

Immediately this vast dominion fell to pieces. One of his generals, Seleucus, retained most of the old Persian empire from the Indus to Ephesus; another, Ptolemy, seized Egypt, and Antigonus secured Macedonia. The rest of the empire remained unstable, passing under the control of a succession of local adventurers. Barbarian raids began from the north and grew in scope and intensity. Until at last, as we shall tell, a new power, the power of the Roman republic, came out of the west to subjugate one fragment after another and weld them together into a new and more enduring empire.

XXVII

The Museum and Library at Alexandria

BEFORE the time of Alexander Greeks had already been spreading as merchants, artists, officials, mercenary soldiers, over most of the Persian dominions. In the dynastic disputes that followed the death of Xerxes, a band of ten thousand Greek mercenaries played a part under the leadership of Xenophon. Their return to Asiatic Greece from Babylon is described in his *Retreat of the Ten Thousand*, one of the first war stories that was ever written by a general in command. But the conquests of Alexander and the division of his brief empire among his subordinate generals, greatly stimulated this permeation of the ancient world by the Greeks and their language and fashions and culture. Traces of this Greek dissemination are to be found far away in central Asia and in north-west India. Their influence upon the development of Indian art was profound.

For many centuries Athens retained her prestige as a centre of art and culture; her schools went on indeed to 529 A.D., that is to say for nearly a thousand years; but the leadership in the intellectual activity of the world passed presently across the Mediterranean to Alexandria, the new trading city that Alexander had founded. Here the Macedonian general Ptolemy had become Pharaoh, with a court that spoke Greek. He had become an intimate of Alexander before he became king, and he was deeply saturated with the ideas of Aristotle. He set himself, with great energy and capacity, to organize knowledge and investigation. He also wrote a history of Alexander's campaigns which, unhappily, is lost to the world.

Alexander had already devoted considerable sums to finance the enquiries of Aristotle, but Ptolemy I was the first person to make a permanent endowment of science. He set up a foundation in Alexandria which was formerly dedicated to the Muses, the Museum of Alexandria. For two or three generations the scientific work done at Alexandria was extraordinarily good. Euclid, Eratosthenes who measured the size of the earth and came within fifty miles of its true diameter, Apollonius who wrote on conic sections, Hipparchus who made the first star map and catalogue, and Hero who devised the first steam engine are among the greater stars of an extraordinary constellation of scientific pioneers. Archimedes came from Syracuse to Alexandria to study, and was a frequent correspondent of the Museum. Herophilus was one of the greatest of Greek anatomists, and is said to have practised vivisection.

For a generation or so during the reigns of Ptolemy I and Ptolemy II there was such a blaze of knowledge and discovery at Alexandria as the world was not to see again until the sixteenth century A.D. But it did not continue. There may have been several causes of this decline. Chief among them, the late Professor Mahaffy suggested, was the fact that the Museum was a "royal" college and all its professors and fellows were appointed and paid by Pharaoh. This was all very well when Pharaoh was Ptolemy I, the pupil and friend of Aristotle. But as the dynasty of the Ptolemies went on they became Egyptianized, they fell under the sway of Egyptian priests and Egyptian religious developments, they ceased to follow the work that was done, and their control stifled the spirit of enquiry altogether. The Museum produced little good work after its first century of activity.

Ptolemy I not only sought in the most modern spirit to organize the finding of fresh knowledge. He tried also to set up an encyclopædic storehouse of wisdom in the Library of Alexandria. It was not simply a storehouse, it was also a book-copying and book-selling organization. A great army of copyists was set to work perpetually multiplying copies of books.

Here then we have the definite first opening up of the intellectual process in which we live to-day; here we have the systematic gathering and distribution of knowledge. The foundation of this Museum and Library marks one of the great epochs in the history of mankind. It is the true beginning of Modern History.

Both the work of research and the work of dissemination went on under serious handicaps. One of these was the great social gap that separated the philosopher, who was a gentleman, from the trader and the artisan. There were glass workers and metal workers in abundance in those days, but they were not in mental contact with the thinkers. The glass worker was making the most beautifully coloured beads and phials and so forth, but he never made a Florentine flask or a lens. Clear glass does not seem to have interested him. The metal worker made weapons and jewellery but he never made a chemical balance. The philosopher speculated loftily about atoms and the nature of things, but he had no practical experience of enamels and pigments and philters and so forth. He was not interested in substances. So Alexandria in its brief day of opportunity produced no microscopes and no chemistry. And though Hero invented a steam engine it was never set either to pump or drive a boat or do any useful thing. There were few practical applications of science except in the realm of medicine, and the progress of science was not stimulated and sustained by the interest and excitement of practical applications. There was nothing to keep the work going therefore when the intellectual curiosity of Ptolemy I and Ptol-

Photo: Dr. Singer

ARISTOTLE

From Herculaneum, probably Fourth Century B.C.

emy II was withdrawn. The discoveries of the Museum went on record in obscure manuscripts and never, until the revival of scientific curiosity at the Renascence, reached out to the mass of mankind.

Nor did the Library produce any improvements in book making. That ancient world had no paper made in definite sizes from rag pulp. Paper was a Chinese invention and it did not reach the western world until the ninth century A.D. The only book materials were parchment and strips of the papyrus reed joined edge to edge. These strips were kept on rolls which were very unwieldy to wind to and fro and read, and very inconvenient for reference. It was these things that prevented the development of paged and printed books. Printing itself was known in the world it would seem as early as the Old Stone Age; there were seals in ancient Sumeria; but without abundant paper there was little advantage in printing books, an improvement that may further have been resisted by trades unionism on the part of the copyists employed. Alexandria produced abundant books but not cheap books, and it never spread knowledge into the population of the ancient world below the level of a wealthy and influential class.

STATUETTE OF MAITREYA: THE BUDDHA TO COME

A Græco-Buddhist sculpture of the Third Century A.D.

(From Malakand, N. W. Province, now in the India Museum)

So it was that this blaze of intellectual enterprise never reached beyond a small circle of people in touch with the group of philosophers collected by the first two Ptolemies. It was like the light in a dark lantern which is shut off from the world at large. Within the blaze may be blindingly bright, but nevertheless it is unseen. The rest of the world went on its old ways unaware that the seed of scientific knowledge that was one day to revolutionize it altogether had been sown. Presently a darkness of bigotry fell even upon

Alexandria. Thereafter for a thousand years of darkness the seed that Aristotle had sown lay hidden. Then it stirred and began to germinate. In a few centuries it had become that widespread growth of knowledge and clear ideas that is now changing the whole of human life.

Alexandria was not the only centre of Greek intellectual activity in the third century B.C. There were many other cities that displayed a brilliant intellectual life amidst the disintegrating fragments of the brief empire of Alexander. There was, for example,

India Mus.

THE DEATH OF BUDDHA

Græco-Buddhist carving from Sivat Valley, N. W. Province, probably A.D. 350

the Greek city of Syracuse in Sicily, where thought and science flourished for two centuries; there was Pergamum in Asia Minor, which also had a great library. But this brilliant Hellenic world was now stricken by invasion from the north. New Nordic barbarians, the Gauls, were striking down along the tracks that had once been followed by the ancestors of the Greeks and Phrygians and Macedonians. They raided, shattered and destroyed. And in the wake of the Gauls came a new conquering people out of Italy, the Romans, who gradually subjugated all the western half of the vast realm of Darius and Alexander. They were an able but unimaginative people, preferring law and profit to either science or art.

New invaders were also coming down out of central Asia to shatter and subdue the Seleucid empire and to cut off the western world again from India. These were the Parthians, hosts of mounted bowmen, who treated the Græco-Persian empire of Persepolis and Susa in the third century B.C. in much the same fashion that the Medes and Persians had treated it in the seventh and sixth. And there were now other nomadic peoples also coming out of the north-east, peoples who were not fair and Nordic and Aryan-speaking but yellow-skinned and black-haired and with a Mongolian speech. But of these latter people we shall tell more in a subsequent chapter.

XXVIII

The Life of Gautama Buddha

BUT now we must go back three centuries in our story to tell of a great teacher who came near to revolutionizing the religious thought and feeling of all Asia. This was Gautama Buddha, who taught his disciples at Benares in India about the same time that Isaiah was prophesying among the Jews in Babylon and Heraclitus was carrying on his speculative enquiries into the nature of things at Ephesus. All these men were in the world at the same time, in the sixth century B.C. — unaware of one another.

This sixth century B.C. was indeed one of the most remarkable in all history. Everywhere — for as we shall tell it was also the case in China — men's minds were displaying a new boldness. Everywhere they were waking up out of the traditions of kingships and priests and blood sacrifices and asking the most penetrating questions. It is as if the race had reached a stage of adolescence — after a childhood of twenty thousand years.

The early history of India is still very obscure. Somewhen perhaps about 2000 B.C., an Aryan-speaking people came down from the north-west into India either in one invasion or in a series of invasions; and was able to spread its language and traditions over most of north India. Its peculiar variety of Aryan speech was the Sanskrit. They found a brunette people with a more elaborate civilization and less vigour of will, in possession of the country of the Indus and Ganges. But they do not seem to have mingled with their predecessors as freely as did the Greeks and Persians. They remained aloof. When the past of India becomes dimly visible to the historian, Indian society is already stratified into several layers, with a variable number of sub-divisions, which do not eat together nor intermarry nor associate freely. And throughout history this stratification into *castes* continues. This makes the Indian population something different from the simple, freely inter-breeding European or Mongolian communities. It is really a community of communities.

Siddhattha Gautama was the son of an aristocratic family which ruled a small district on the Himalayan slopes. He was married at nineteen to a beautiful cousin. He hunted and played and went about in his sunny world of gardens and groves and irrigated rice-fields. And it was amidst this life that a great discontent fell upon him. It was the unhappiness of a fine brain that seeks employment. He felt that the existence he was leading was not the reality of life, but a holiday — a holiday that had gone on too long.

The sense of disease and mortality, the insecurity and the unsatisfactoriness of all happiness, descended upon the mind of Gautama. While he was in this mood he met one of those wandering ascetics who already existed in great numbers in India. These men lived under severe rules, spending much time in meditation and in religious discussion. They were supposed to be seeking some deeper reality in life, and a passionate desire to do likewise took possession of Gautama.

He was meditating upon this project, says the story, when the news was brought to him that his wife had been delivered of his first-born son. "This is another tie to break," said Gautama.

He returned to the village amidst the rejoicings of his fellow clansmen. There was a great feast and a Nautch dance to celebrate the birth of this new tie, and in the night Gautama awoke in a great agony of spirit, "like a man who is told that his house is on fire." He resolved to leave his happy aimless life forthwith. He went softly to the threshold of his wife's chamber, and saw her by the light of a little oil lamp, sleeping sweetly, surrounded by flowers, with his infant son in her arms. He felt a great craving to take up the child in one first and last embrace before he departed, but the fear of waking his wife prevented him, and at last he turned away and went out into the bright Indian moonshine and mounted his horse and rode off into the world.

Very far he rode that night, and in the morning he stopped out-

India Mus.

TIBETAN BUDDHA

Gilt Brass Casting in India Museum, showing Gautama Buddha in the "earth witness" attitude

side the lands of his clan, and dismounted beside a sandy river. There he cut off his flowing locks with his sword, removed all his ornaments and sent them and his horse and sword back to his house. Going on he presently met a ragged man and exchanged clothes with him, and so having divested himself of all worldly entanglements he was free to pursue his search after wisdom. He made his way southward to a resort of hermits and teachers in a hilly spur of the Vindhya Mountains. There lived a number of wise men in a warren of caves, going into the town for their simple supplies and imparting their knowledge by word of mouth to such as cared to come to them. Gautama became versed in all the metaphysics of his age. But his acute intelligence was dissatisfied with the solutions offered him.

The Indian mind has always been disposed to believe that power and knowledge may be obtained by extreme asceticism, by fasting, sleeplessness, and self-torment, and these ideas Gautama now put to the test. He betook himself with five disciple companions to the jungle and there he gave himself up to fasting and terrible penances. His fame spread, "like the sound of a great

A BURMESE BUDDHA

Marble Figure from Mandalay, eighteenth century work, now in the India Museum

THE DHAMÊKH TOWER
In the Deer Park at Sarnath. Sixth Century A.D.
(From a Painting in the India Museum)

bell hung in the canopy of the skies." But it brought him no sense of truth achieved. One day he was walking up and down, trying to think in spite of his enfeebled state. Suddenly he fell unconscious. When he recovered, the preposterousness of these semi-magical ways to wisdom was plain to him.

He horrified his companions by demanding ordinary food and refusing to continue his mortifications. He had realized that whatever truth a man may reach is reached best by a nourished brain in a healthy body. Such a conception was absolutely foreign to the ideas of the land and age. His disciples deserted him, and went off in a melancholy state to Benares. Gautama wandered alone.

When the mind grapples with a great and intricate problem, it makes its advances step by step, with but little realization of the gains it has made, until suddenly, with an effect of abrupt illumination, it realizes its victory. So it happened to Gautama. He had seated himself under a great tree by the side of a river to eat, when this sense of clear vision came to him. It seemed to him that he saw life plain. He is said to have sat all day and all night in profound thought, and then he rose up to impart his vision to the world.

He went on to Benares and there he sought out and won back his lost disciples to his new teaching. In the King's Deer Park at Benares they built themselves huts and set up a sort of school to which came many who were seeking after wisdom.

The starting point of his teaching was his own question as a fortunate young man, "Why am I not completely happy?" It was an introspective question. It was a question very different in quality from the frank and self-forgetful *externalized* curiosity with which Thales and Heraclitus were attacking the problems of the universe, or the equally self-forgetful burthen of moral obligation that the culminating prophets were imposing upon the Hebrew mind. The Indian teacher did not forget self, he concentrated upon self and sought to destroy it. All suffering, he taught, was due to the greedy desires of the individual. Until man has conquered his personal cravings his life is trouble and his end sorrow. There were three principal forms that the craving for life took and they were all evil. The first was the desire of the appetites, greed and all forms of sensuousness, the second was the desire for a personal and egotistic immortality, the third was the craving for personal success, worldliness, avarice and the like. All these forms of desire had to be overcome to escape from the distresses and chagrins of life. When they were overcome, when self had vanished altogether, then serenity of soul, Nirvana, the highest good was attained.

This was the gist of his teaching, a very subtle and metaphysical teaching indeed, not nearly so easy to understand as the Greek injunction to see and know fearlessly and rightly and the Hebrew command to fear God and accomplish righteousness. It was a teaching much beyond the understanding of even Gautama's immediate disciples, and it is no wonder that so soon as his personal influence was withdrawn it became corrupted and coarsened. There was a widespread belief in India at that time that at long intervals Wisdom came to earth and was incarnate in some chosen person who was known as the Buddha. Gautama's disciples declared that he was a Buddha, the latest of the Buddhas, though there is no evidence that he himself ever accepted the title. Before he was well dead, a cycle of fantastic legends began to be woven about him. The human heart has always preferred a wonder story to a moral effort, and Gautama Buddha became very wonderful.

Yet there remained a substantial gain in the world. If Nirvana was too high and subtle for most men's imaginations, if the myth-making impulse in the race was too strong for the simple facts of Gautama's life, they could at least grasp something of the intention of what Gautama called the Eight-fold way, the Aryan or Noble Path in life. In this there was an insistence upon mental uprightness, upon right aims and speech, right conduct and honest livelihood. There was a quickening of the conscience and an appeal to generous and self-forgetful ends.

XXIX

King Asoka

FOR some generations after the death of Gautama, these high and noble Buddhist teachings, this first plain teaching that the highest good for man is the subjugation of self, made comparatively little headway in the world. Then they conquered the imagination of one of the greatest monarchs the world has ever seen.

We have already mentioned how Alexander the Great came down into India and fought with Porus upon the Indus. It is related by the Greek historians that a certain Chandragupta Maurya came into Alexander's camp and tried to persuade him to go on to the Ganges and conquer all India. Alexander could not do this because of the refusal of his Macedonians to go further into what was for them an unknown world, and later on (321 B.C.) Chandragupta was able to secure the help of various hill tribes and realize his dream without Greek help. He built up an empire in North India and was presently (303 B.C.) able to attack Seleucus I in the Punjab and drive the last vestige of Greek power out of India. His son extended this new empire. His grandson, Asoka, the monarch of whom we now have to tell, found himself in 264 B.C. ruling from Afghanistan to Madras.

Asoka was at first disposed to follow the example of his father and grandfather and complete the conquest of the Indian peninsula. He invaded Kalinga (255 B.C.), a country on the east coast of Madras, he was successful in his military operations and — alone among conquerors — he was so disgusted by the cruelty and horror of war that he renounced it. He would have no more of it. He adopted the peaceful doctrines of Buddhism and declared that henceforth his conquests should be the conquests of religion.

His reign for eight-and-twenty years was one of the brightest interludes in the troubled history of mankind. He organized a great digging of wells in India and the planting of trees for shade. He founded hospitals and public gardens and gardens for the growing of medicinal herbs. He created a ministry for the care of the

A LOHAN OR BUDDHIST APOSTLE (Tang Dynasty)
(From the statue in the British Museum)

aborigines and subject races of India. He made provision for the education of women. He made vast benefactions to the Buddhist teaching orders, and tried to stimulate them to a better and more energetic criticism of their own accumulated literature. For corruptions and superstitious accretions had accumulated very

India Mus.

TRANSOME SHOWING THE COURT OF ASOKA

speedily upon the pure and simple teaching of the great Indian master. Missionaries went from Asoka to Kashmir, to Persia, to Ceylon and Alexandria.

Such was Asoka, greatest of kings. He was far in advance of his age. He left no prince and no organization of men to carry on his work, and within a century of his death the great days of his reign had become a glorious memory in a shattered and decaying India. The priestly caste of the Brahmins, the highest and most privileged caste in the Indian social body, has always been opposed to the frank and open teaching of Buddha. Gradually they undermined the Buddhist influence in the land. The old monstrous gods, the innumerable cults of Hinduism, resumed their sway. Caste became

India Mus.

ASOKA PANEL FROM BHARHUT

THE PILLAR OF LIONS

Capital of the Pillar (column lying on side) erected in Deer Park in the time of Asoka, where Buddha preached his first sermon

(*From a print in the India Museum*)

more rigorous and complicated. For long centuries Buddhism and Brahminism flourished side by side, and then slowly Buddhism decayed and Brahminism in a multitude of forms replaced it. But beyond the confines of India and the realms of caste Buddhism spread — until it had won China and Siam and Burma and Japan, countries in which it is predominant to this day.

XXX

Confucius and Lao Tse

WE have still to tell of two other great men, Confucius and Lao Tse, who lived in that wonderful century which began the adolescence of mankind, the sixth century B.C.

In this history thus far we have told very little of the early story of China. At present that early history is still very obscure, and we look to Chinese explorers and archæologists in the new China that is now arising to work out their past as thoroughly as the European past has been worked out during the last century. Very long ago the first primitive Chinese civilizations arose in the great river valleys out of the primordial heliolithic culture. They had, like Egypt and Sumeria, the general characteristics of that culture, and they centred upon temples in which priests and priest kings offered the seasonal blood sacrifices. The life in those cities must have been very like the Egyptian and Sumerian life of six or seven thousand years ago and very like the Maya life of Central America a thousand years ago.

If there were human sacrifices they had long given way to animal sacrifices before the dawn of history. And a form of picture writing was growing up long before a thousand years B.C.

And just as the primitive civilizations of Europe and western Asia were in conflict with the nomads of the desert and the nomads of the north, so the primitive Chinese civilizations had a great cloud of nomadic peoples on their northern borders. There was a number of tribes akin in language and ways of living, who are spoken of in history in succession as the Huns, the Mongols, the Turks and Tartars. They changed and divided and combined and re-combined, just as the Nordic peoples in north Europe and central Asia changed and varied in name rather than in nature. These Mongolian nomads had horses earlier than the Nordic peoples, and it may be that in the region of the Altai Mountains they made an independent discovery of iron somewhen after 1000 B.C. And just as in the western case so ever and again these eastern nomads would achieve a sort of political unity, and become the conquerors and masters and revivers of this or that settled and civilized region.

It is quite possible that the earliest civilization of China was not Mongolian at all any more than the earliest civilization of Europe and western Asia was Nordic or Semitic. It is quite possible that the earliest civilization of China was a brunette civilization and of a piece with the earliest Egyptian, Sumerian and Dravidian civilizations, and that when the first recorded history of China began there had already been conquests and intermixture. At any rate we find that by 1750 B.C. China was already a vast system of little kingdoms and city states, all acknowledging a loose allegiance and paying more or less regularly, more or less definite feudal dues to one great priest emperor, the "Son of Heaven." The "Shang" dynasty came to an end in 1125 B.C. A "Chow" dynasty succeeded "Shang," and maintained China in a relaxing unity until the days of Asoka in India and of the Ptolemies in Egypt. Gradually China went to pieces during that long "Chow" period. Hunnish peoples came down and set up principalities; local rulers discontinued their tribute and became independent. There was in the sixth century B.C., says one Chinese authority, five or six thousand practically independent states in China. It was what the Chinese call in their records an "Age of Confusion."

But this Age of Confusion was compatible with much intellectual activity and with the existence of many local centres of art and civilized living. When we know more of Chinese history we shall find that China also had her Miletus and her Athens, her Pergamum and her Macedonia. At present we must be vague and brief about this period of Chinese division simply because our knowledge is not sufficient for us to frame a coherent and consecutive story.

And just as in divided Greece there were philosophers and in shattered and captive Jewry prophets, so in disordered China there were philosophers and teachers at this time. In all these cases

CONFUCIUS

Copy of stone carving in the Temple of Confucius at K'iu Fu

(From the records of the Archæological Mission to North China (Chavannes)

insecurity and uncertainty seemed to have quickened the better sort of mind. Confucius was a man of aristocratic origin and some official importance in a small state called Lu. Here in a very parallel mood to the Greek impulse he set up a sort of Academy for discovering and teaching Wisdom. The lawlessness and disorder of China distressed him profoundly. He conceived an ideal of a better government and a better life, and travelled from state to state seeking a prince who would carry out his legislative and educational ideas. He never found his prince; he found a prince, but court intrigues undermined the influence of the teacher and finally defeated his reforming proposals. It is interesting to note that a century and a half later the Greek philosopher Plato also sought a prince, and was for a time adviser to the tyrant Dionysius who ruled Syracuse in Sicily.

Confucius died a disappointed man. "No intelligent ruler arises to take me as his master," he said, "and my time has come to die." But his teaching had more vitality than he imagined in his declining and hopeless years, and it became a great formative influence with the Chinese people. It became one of what the Chinese call the Three Teachings, the other two being those of Buddha and of Lao Tse.

The gist of the teaching of Confucius was the way of the noble or aristocratic man. He was concerned with personal conduct as much as Gautama was concerned with the peace of self-forgetfulness and the Greek with external knowledge and the Jew with righteousness. He was the most public-minded of all great teachers. He was supremely concerned by the confusion and miseries of the world, and he wanted to make men noble in order to bring about a noble world. He sought to regulate conduct to an extraordinary extent; to provide sound rules for every occasion in life. A polite, public-spirited gentleman, rather sternly self-disciplined, was the ideal he found already developing in the northern Chinese world and one to which he gave a permanent form.

The teaching of Lao Tse, who was for a long time in charge of the imperial library of the Chow dynasty, was much more mystical and vague and elusive than that of Confucius. He seems to have preached a stoical indifference to the pleasures and powers of the

Photo: Underwood & Underwood

THE GREAT WALL OF CHINA

As it crosses the mountains in Manchuria

world and a return to an imaginary simple life of the past. He left writings very contracted in style and very obscure. He wrote in riddles. After his death his teachings, like the teachings of Gautama Buddha, were corrupted and overlaid by legends and had the most complex and extraordinary observances and superstitious ideas grafted upon them. In China just as in India primordial ideas of magic and monstrous legends out of the childish past of our race struggled against the new thinking in the world and succeeded in plastering it over with grotesque, irrational and antiquated observances. Both Buddhism and Taoism (which ascribes itself largely to Lao Tse) as one finds them in China now, are religions of monk, temple, priest and offering of a type as ancient in form, if not in thought, as the sacrificial religions of ancient Sumeria and Egypt. But the teaching

EARLY CHINESE BRONZE BELL

Inscribed in archaic characters: "made for use by the elder of Hing village in Ting district;" latter half of the Chou Dynasty, Sixth Century B.C.

(In the Victoria and Albert Museum)

of Confucius was not so overlaid because it was limited and plain and straightforward and lent itself to no such distortions.

North China, the China of the Hwang-ho River, became Confucian in thought and spirit; south China, Yang-tse-Kiang China, became Taoist. Since those days a conflict has always been traceable in Chinese affairs between these two spirits, the spirit of the north and the spirit of the south, between (in latter times) Pekin and Nankin, between the official-minded, upright and conservative north, and the sceptical, artistic, lax and experimental south.

The divisions of China of the Age of Confusion reached their worst stage in the sixth century B.C. The Chow dynasty was so enfeebled and so discredited that Lao Tse left the unhappy court and retired into private life.

Three nominally subordinate powers dominated the situation in those days, Ts'i and Ts'in, both northern powers, and Ch'u, which was an aggressive military power in the Yangtse valley. At last Ts'i and Ts'in formed an alliance, subdued Ch'u and imposed a general treaty of disarmament and peace in China. The power of Ts'in became predominant. Finally about the time of Asoka in India the Ts'in monarch seized upon the sacrificial vessels of the Chow emperor and took over his sacrificial duties. His son, Shi-Hwang-ti (king in 246 B.C., emperor in 220 B.C.), is called in the Chinese Chronicles "the First Universal Emperor."

More fortunate than Alexander, Shi-Hwang-ti reigned for thirty-six years as king and emperor. His energetic reign marks the beginning of a new era of unity and prosperity for the Chinese people. He fought vigorously against the Hunnish invaders from the northern deserts, and he began that immense work, the Great Wall of China, to set a limit to their incursions.

XXXI

Rome Comes into History

THE reader will note a general similarity in the history of all these civilizations in spite of the effectual separation caused by the great barriers of the Indian north-west frontier and of the mountain masses of Central Asia and further India. First for thousands of years the heliolithic culture spread over all the warm and fertile river valleys of the old world and developed a temple system and priest rulers about its sacrificial traditions. Apparently its first makers were always those brunette peoples we have spoken of as the central race of mankind. Then the nomads came in from the regions of seasonal grass and seasonal migrations and superposed their own characteristics and often their own language on the primitive civilization. They subjugated and stimulated it, and were stimulated to fresh developments and made it here one thing and here another. In Mesopotamia it was the Elamite and then the Semite, and at last the Nordic Medes and Persians and the Greeks who supplied the ferment; over the region of the Ægean peoples it was the Greeks; in India it was the Aryan-speakers; in Egypt there was a thinner infusion of conquerors into a more intensely saturated priestly civilization; in China, the Hun conquered and was absorbed and was followed by fresh Huns. China was Mongolized just as Greece and North India were Aryanized and Mesopotamia Semitized and Aryanized. Everywhere the nomads destroyed much, but everywhere they brought in a new spirit of free enquiry and moral innovation. They questioned the beliefs of immemorial ages. They let daylight into the temples. They set up kings who were neither priests nor gods but mere leaders among their captains and companions.

In the centuries following the sixth century B.C. we find everywhere a great breaking down of ancient traditions and a new spirit

Photo: Anderson

THE DYING GAUL

The statue in the National Museum, Rome, depicting a Gaul stabbing himself, after killing his wife, in the presence of his enemies

of moral and intellectual enquiry awake, a spirit never more to be altogether stilled in the great progressive movement of mankind. We find reading and writing becoming common and accessible accomplishments among the ruling and prosperous minority; they were no longer the jealously guarded secret of the priests. Travel is increasing and transport growing easier by reason of horses and roads. A new and easy device to facilitate trade has been found in coined money.

Let us now transfer our attention back from China in the extreme east of the old world to the western half of the Mediterranean. Here we have to note the appearance of a city which was destined to play at last a very great part indeed in human affairs, Rome.

Hitherto we have told very little about Italy in our story. It was before 1000 B.C. a land of mountain and forest and thinly populated. Aryan-speaking tribes had pressed down this peninsula and formed little towns and cities, and the southern extremity was studded with Greek settlements. The noble ruins of Pæstum preserve for us to this day something of the dignity and splendour of these early Greek establishments. A non-Aryan people, probably akin to the Ægean peoples, the Etruscans, had established themselves in the central part of the peninsula. They had reversed the usual process by subjugating various Aryan tribes. Rome, when it comes into the light of history, is a little trading city at a ford on the Tiber, with a Latin-speaking population ruled over by Etruscan kings. The old chronologies gave 753 B.C. as the date of the founding of Rome, half a century later than the founding of the great Phœnician city of Carthage and twenty-three years after the first Olympiad. Etruscan tombs of a much earlier date than 753 B.C. have, however, been excavated in the Roman Forum.

In that red-letter century, the sixth century B.C., the Etruscan kings were expelled (510 B.C.) and Rome became an aristocratic republic with a lordly class of "patrician" families dominating a commonalty of "plebeians." Except that it spoke Latin it was not unlike many aristocratic Greek republics.

For some centuries the internal history of Rome was the story of a long and obstinate struggle for freedom and a share in the government on the part of the plebeians. It would not be difficult to find

Greek parallels to this conflict, which the Greeks would have called a conflict of aristocracy with democracy. In the end the plebeians broke down most of the exclusive barriers of the old families and established a working equality with them. They destroyed the old exclusiveness, and made it possible and acceptable for Rome to extend her citizenship by the inclusion of more and more "outsiders." For while she still struggled at home, she was extending her power abroad.

Photo: Underwood & Underwood

REMAINS OF THE ANCIENT ROMAN CISTERNS AT CARTHAGE

The extension of Roman power began in the fifth century B.C. Until that time they had waged war, and generally unsuccessful war, with the Etruscans. There was an Etruscan fort, Veii, only a few miles from Rome which the Romans had never been able to capture. In 474 B.C., however, a great misfortune came to the Etruscans. Their fleet was destroyed by the Greeks of Syracuse in Sicily.

At the same time a wave of Nordic invaders came down upon them from the north, the Gauls. Caught between Roman and Gaul, the Etruscans fell — and disappear from history. Veii was captured by the Romans. The Gauls came through to Rome and sacked the city (390 B.C.) but could not capture the Capitol. An attempted night surprise was betrayed by the cackling of some geese, and finally the invaders were bought off and retired to the north of Italy again.

The Gaulish raid seems to have invigorated rather than weakened Rome. The Romans conquered and assimilated the Etruscans, and extended their power over all central Italy from the Arno to Naples. To this they had reached within a few years of 300 B.C. Their conquests in Italy were going on simultaneously with the growth of Philip's power in Macedonia and Greece, and the tremendous raid of Alexander to Egypt and the Indus. The Romans had become notable people in the civilized world to the east of them by the break-up of Alexander's empire.

To the north of the Roman power were the Gauls; to the south of them were the Greek settlements of Magna Græcia, that is to say of Sicily and of the toe and heel of Italy. The Gauls were a hardy, warlike people and the Romans held that boundary by a line of forts and fortified settlements. The Greek cities in the south headed by Tarentum (now Taranto) and by Syracuse in Sicily, did not so much threaten as fear the Romans. They looked about for some help against these new conquerors.

We have already told how the empire of Alexander fell to pieces and was divided among his generals and companions. Among these adventurers was a kinsman of Alexander's named Pyrrhus, who established himself in Epirus, which is across the Adriatic Sea over against the heel of Italy. It was his ambition to play the part of Philip of Macedonia to Magna Græcia, and to become protector and master-general of Tarentum, Syracuse and the rest of that part of the world. He had what was then a very efficient modern army; he had an infantry phalanx, cavalry from Thessaly — which was now quite as good as the original Macedonian cavalry — and twenty fighting elephants; he invaded Italy and routed the Romans in two considerable battles, Heraclea (280 B.C.) and Ausculum (279 B.C.), and having driven them north, he turned his attention to the subjugation of Sicily.

But this brought against him a more formidable enemy than were the Romans at that time, the Phœnician trading city of Carthage, which was probably then the greatest city in the world. Sicily was too near Carthage for a new Alexander to be welcome there, and Carthage was mindful of the fate that had befallen her mother city Tyre half a century before. So she sent a fleet to encourage or compel Rome to continue the struggle, and she cut the overseas communications of Pyrrhus. Pyrrhus found himself freshly assailed by the Romans, and suffered a disastrous repulse in an attack he had made upon their camp at Beneventum between Naples and Rome.

And suddenly came news that recalled him to Epirus. The Gauls were raiding south. But this time they were not raiding down into Italy; the Roman frontier, fortified and guarded, had become too formidable for them. They were raiding down through Illyria (which is now Serbia and Albania) to Macedonia and Epirus. Repulsed by the Romans, endangered at sea by the Carthaginians, and threatened at home by the Gauls, Pyrrhus abandoned his dream of conquest and went home (275 B.C.), and the power of Rome was extended to the Straits of Messina.

On the Sicilian side of the Straits was the Greek city of Messina, and this presently fell into the hands of a gang of pirates. The Carthaginians, who were already practically overlords of Sicily and allies of Syracuse, suppressed these pirates (270 B.C.) and put in a Carthaginian garrison there. The pirates appealed to Rome and Rome listened to their complaint. And so across the Straits of Messina the great trading power of Carthage and this new conquering people, the Romans, found themselves in antagonism, face to face.

XXXII

Rome and Carthage

IT was in 264 B.C. that the great struggle between Rome and Carthage, the Punic Wars, began. In that year Asoka was beginning his reign in Behar and Shi-Hwang-ti was a little child, the Museum in Alexandria was still doing good scientific work, and the barbaric Gauls were now in Asia Minor and exacting a tribute from Pergamum. The different regions of the world were still separated by insurmountable distances, and probably the rest of mankind heard only vague and remote rumours of the mortal fight that went on for a century and a half in Spain, Italy, North Africa and the western Mediterranean, between the last stronghold of Semitic power and Rome, this newcomer among Aryan-speaking peoples.

That war has left its traces upon issues that still stir the world. Rome triumphed over Carthage, but the rivalry of Aryan and Semite was to merge itself later on in the conflict of Gentile and Jew. Our history now is coming to events whose consequences and distorted traditions still maintain a lingering and expiring vitality in, and exercise a complicating and confusing influence upon, the conflicts and controversies of to-day.

The First Punic War began in 264 B.C. about the pirates of Messina. It developed into a struggle for the possession of all Sicily except the dominions of the Greek king of Syracuse. The advantage of the sea was at first with the Carthaginians. They had great fighting ships of what was hitherto an unheard-of size, quinqueremes, galleys with five banks of oars and a huge ram. At the battle of Salamis, two centuries before, the leading battleships had only been triremes with three banks. But the Romans, with extraordinary energy and in spite of the fact that they had little naval experience, set themselves to outbuild the Carthaginians. They manned the new navy they created chiefly with Greek seamen, and they invented grappling and boarding to make up for the superior seamanship of the enemy. When the Carthaginian came up to ram or shear the oars of the Roman, huge grappling irons seized him and the Roman soldiers swarmed aboard him. At Mylæ (260 B.C.) and at Ecnomus

Photo: Mansell

HANNIBAL
Bust in the National Museum at Naples

(256 B.C.) the Carthaginians were disastrously beaten. They repulsed a Roman landing near Carthage but were badly beaten at Palermo, losing one hundred and four elephants there — to grace such a triumphal procession through the Forum as Rome had never seen before. But after that came two Roman defeats and then a Roman recovery. The last naval forces of Carthage were defeated

by a last Roman effort at the battle of the Ægatian Isles (241 B.C.) and Carthage sued for peace. All Sicily except the dominions of Hiero, king of Syracuse, was ceded to the Romans.

For twenty-two years Rome and Carthage kept the peace. Both had trouble enough at home. In Italy the Gauls came south again, threatened Rome — *which in a state of panic offered human sacrifices to the Gods!* — and were routed at Telamon. Rome pushed forward to the Alps, and even extended her dominions down the Adriatic coast to Illyria. Carthage suffered from domestic insurrections and from revolts in Corsica and Sardinia, and displayed far less recuperative power. Finally, an act of intolerable aggression, Rome seized and annexed the two revolting islands.

Spain at that time was Carthaginian as far north as the river Ebro. To that boundary the Romans restricted them. Any crossing of the Ebro by the Carthaginians was to be considered an act of war against the Romans. At last in 218 B.C. the Carthaginians, provoked by new Roman aggressions, did cross this river under a young general named Hannibal, one of the most brilliant commanders in the whole of history. He marched his army from Spain over the Alps into Italy, raised the Gauls against the Romans, and carried on the Second Punic War in Italy itself for fifteen years. He inflicted tremendous defeats upon the Romans at Lake Trasimere and at Cannæ, and throughout all his Italian campaigns no Roman army stood against him and escaped disaster. But a Roman army had landed at Marseilles and cut his communications with Spain; he had no siege train, and he could never capture Rome. Finally the Carthaginians, threatened by the revolt of the Numidians at home, were forced back upon the defence of their own city in Africa, a Roman army crossed into Africa, and Hannibal experienced his first defeat under its walls at the battle of Zama (202 B.C.) at the hands of Scipio Africanus the Elder. The battle of Zama ended this Second Punic War. Carthage capitulated; she surrendered Spain and her war fleet; she paid an enormous indemnity and agreed to give up Hannibal to the vengeance of the Romans. But Hannibal escaped and fled to Asia where later, being in danger of falling into the hands of his relentless enemies, he took poison and died.

For fifty-six years Rome and the shorn city of Carthage were at peace. And meanwhile Rome spread her empire over confused and divided Greece, invaded Asia Minor, and defeated Antiochus III, the Seleucid monarch, at Magnesia in Lydia. She made Egypt, still under the Ptolemies, and Pergamum and most of the small states of Asia Minor into "Allies," or, as we should call them now, "protected states."

Meanwhile Carthage, subjugated and enfeebled, had been slowly regaining something of her former prosperity. Her recovery revived the hate and suspicion of the Romans. She was attacked upon the most shallow and artificial of quarrels (149 B.C.), she made an obstinate and bitter resistance, stood a long siege and was stormed (146 B.C.). The street fighting, or massacre, lasted six days; it was extraordinarily bloody, and when the citadel capitulated only about fifty thousand of the Carthaginian population remained alive out of a quarter of a million. They were sold into slavery, and the city was burnt and elaborately destroyed. The blackened ruins were ploughed and sown as a sort of ceremonial effacement.

So ended the Third Punic War. Of all the Semitic states and cities that had flourished in the world five centuries before only one little country remained free under native rulers. This was Judea, which had liberated itself from the Seleucids and was under the rule

of the native Maccabean princes. By this time it had its Bible almost complete, and was developing the distinctive traditions of the Jewish world as we know it now. It was natural that the Carthaginians, Phœnicians and kindred peoples dispersed about the world should find a common link in their practically identical language and in this literature of hope and courage. To a large extent they were still the traders and bankers of the world. The Semitic world had been submerged rather than replaced.

Jerusalem, which has always been rather the symbol than the centre of Judaism, was taken by the Romans in 65 B.C.; and after various vicissitudes of quasi-independence and revolt was besieged by them in 70 A.D. and captured after a stubborn struggle. The Temple was destroyed. A later rebellion in 132 A.D. completed its destruction, and the Jerusalem we know to-day was rebuilt later under Roman auspices. A temple to the Roman god, Jupiter Capitolinus, stood in the place of the Temple, and Jews were forbidden to inhabit the city.

XXXIII

The Growth of the Roman Empire

NOW this new Roman power which arose to dominate the western world in the second and first centuries B.C. was in several respects a different thing from any of the great empires that had hitherto prevailed in the civilized world. It was not at first a monarchy, and it was not the creation of any one great conqueror. It was not indeed the first of republican empires; Athens had dominated a group of Allies and dependents in the time of Pericles, and Carthage when she entered upon her fatal struggle with Rome was mistress of Sardinia and Corsica, Morocco, Algiers, Tunis, and most of Spain and Sicily. But it was the first republican empire that escaped extinction and went on to fresh developments.

The centre of this new system lay far to the west of the more ancient centres of empire, which had hitherto been the river valleys of Mesopotamia and Egypt. This westward position enabled Rome to bring in to civilization quite fresh regions and peoples. The Roman power extended to Morocco and Spain, and was presently able to thrust north-westward over what is now France and Belgium to Britain and north-eastward into Hungary and South Russia. But on the other hand it was never able to maintain itself in Central Asia or Persia because they were too far from its administrative centres. It included therefore great masses of fresh Nordic Aryan-speaking peoples, it presently incorporated nearly all the Greek people in the world, and its population was less strongly Hamitic and Semitic than that of any preceding empire.

For some centuries this Roman Empire did not fall into the grooves of precedent that had so speedily swallowed up Persian and Greek, and all that time it developed. The rulers of the Medes and Persians became entirely Babylonized in a generation or so; they took over the tiara of the king of kings and the temples and priesthoods of his gods; Alexander and his successors followed in the same easy path of assimilation; the Seleucid monarchs had much the same court and administrative methods as Nebuchadnezzar; the Ptolemies became Pharaohs and altogether Egyptian. They were assimilated just as before them the Semitic conquerors of the Sumerians had been assimilated. But the Romans ruled in their own city, and for some centuries kept to the laws of their own nature. The only people who exercised any great mental influence upon them before the second or third century A.D. were the kindred and similar Greeks. So that the Roman Empire was essentially a first attempt to rule a great dominion upon mainly Aryan lines. It was so far a new pattern in history, it was an expanded Aryan republic. The old pattern of a personal conqueror ruling over a capital city that had grown up round the temple of a harvest god did not apply to it. The Romans had gods and temples, but like the gods of the Greeks their gods were quasi-human immortals, divine patricians. The Romans also had blood sacrifices and even made human ones in times of stress, things they may have learnt to do from their dusky Etruscan teachers; but until Rome was long past its zenith neither priest nor temple played a large part in Roman history.

The Roman Empire was a growth, an unplanned novel growth; the Roman people found themselves engaged almost unawares in a vast administrative experiment. It cannot be called a successful experiment. In the end their empire collapsed altogether. And it changed enormously in form and method from century to century. It changed more in a hundred years than Bengal or Mesopotamia or Egypt changed in a thousand. It was always changing. It never attained to any fixity.

In a sense the experiment failed. In a sense the experiment remains unfinished, and Europe and America to-day are still working out the riddles of world-wide statescraft first confronted by the Roman people.

It is well for the student of history to bear in mind the very great changes not only in political but in social and moral matters that went on throughout the period of Roman dominion. There is much too strong a tendency in people's minds to think of the Roman rule as something finished and stable, firm, rounded, noble and decisive. Macaulay's *Lays of Ancient Rome*, S.P.Q.R. the elder Cato, the Scipios, Julius Cæsar, Diocletian, Constantine the Great, triumphs, orations, gladiatorial combats and Christian martyrs are all mixed up together in a picture of something high and cruel and dignified. The items of that picture have to be disentangled. They are collected at different points from a process of change profounder than that which separates the London of William the Conqueror from the London of to-day.

We may very conveniently divide the expansion of Rome into four stages. The first stage began after the sack of Rome by the Goths in 390 B.C. and went on until the end of the First Punic War (240 B.C.). We may call this stage the stage of the Assimilative Republic. It was perhaps the finest, most characteristic stage in Roman history. The age-long dissensions of patrician and plebeian were drawing to a close, the Etruscan threat had come to an end, no one was very rich yet nor very poor, and most men were public-spirited. It was a republic like the republic of the South African Boers before 1900 or like the northern states of the American Union between 1800 and 1850; a free-farmers republic. At the outset of this stage Rome was a little state scarcely twenty miles square. She fought the sturdy but kindred states about her, and sought not their destruction but coalescence. Her centuries of civil dissension had trained her people in compromise and concessions. Some of the defeated cities became altogether Roman with a voting share in the government, some became self-governing with the right to trade and marry in Rome; garrisons full of citizens were set up at strategic points and colonies of varied privileges founded among the freshly conquered people. Great roads were made. The rapid Latinization of all Italy was the inevitable consequence of such a policy. In 89 B.C. all the free inhabitants of Italy became citizens of the city of Rome Formally the whole Roman Empire became at last an extended city. In 212 A.D. every free man in the entire extent of the empire was given citizenship; the right, if he could get there, to vote in the town meeting in Rome.

This extension of citizenship to tractable cities and to whole countries was the distinctive device of Roman expansion. It reversed the old process of conquest and assimilation altogether. By the Roman method the conquerors assimilated the conquered.

But after the First Punic War and the annexation of Sicily, though the old process of assimilation still went on, another process arose by its side. Sicily for instance was treated as a conquered prey. It was declared an "estate" of the Roman people. Its rich soil and industrious population was exploited to make Rome rich. The patricians and the more influential among the plebeians secured

THE FORUM AT ROME AS IT IS TO-DAY

the major share of that wealth. And the war also brought in a large supply of slaves. Before the First Punic War the population of the republic had been largely a population of citizen farmers. Military service was their privilege and liability. While they were on active service their farms fell into debt and a new large-scale slave agriculture grew up; when they returned they found their produce in competition with slave-grown produce from Sicily and from the new estates at home. Times had changed. The republic had altered its character. Not only was Sicily in the hands of Rome, the common man was in the hands of the rich creditor and the rich competitor. Rome had entered upon its second stage, the Republic of Adventurous Rich Men.

For two hundred years the Roman soldier farmers had struggled for freedom and a share in the government of their state; for a

Photo: Jacques Boyer

RELICS OF ROMAN RULE

Ruins of Coliseum in Tunis

hundred years they had enjoyed their privileges. The First Punic War wasted them and robbed them of all they had won.

The value of their electoral privileges had also evaporated. The governing bodies of the Roman republic were two in number. The first and more important was the Senate. This was a body originally of patricians and then of prominent men of all sorts, who were summoned to it first by certain powerful officials, the consuls and censors. Like the British House of Lords it became a gathering of great landowners, prominent politicians, big business men and the like. It was much more like the British House of Lords than it was like the American Senate. For three centuries, from the Punic Wars onward, it was the centre of Roman political thought and purpose. The second body was the Popular Assembly. This was supposed to be an assembly of *all* the citizens of Rome. When Rome was a little state twenty miles square this was a possible gathering. When the citizenship of Rome had spread beyond the confines in Italy, it was an altogether impossible one. Its meetings, proclaimed by horn-blowing from the Capitol and the city walls, became more and more a gathering of political hacks and city riff-raff. In the fourth century B.C. the Popular Assembly was a considerable check upon the Senate, a competent representation of the claims and rights of the common man. By the end of the Punic Wars it was an impotent relic of a vanquished popular control. No effectual legal check remained upon the big men.

Nothing of the nature of representative government was ever introduced into the Roman republic. No one thought of electing delegates to represent the will of the citizens. This is a very important point for the student to grasp. The Popular Assembly

never became the equivalent of the American House of Representatives or the British House of Commons. In theory it was all the citizens; in practice it ceased to be anything at all worth consideration.

The common citizen of the Roman Empire was therefore in a very poor case after the Second Punic War; he was impoverished, he had often lost his farm, he was ousted from profitable production by slaves, and he had no political power left to him to remedy these things. The only methods of popular expression left to a people without any form of political expression are the strike and the revolt. The story of the second and first centuries B.C., so far as internal politics go, is a story of futile revolutionary upheaval. The scale of this history will not permit us to tell of the intricate struggles of that time, of the attempts to break up estates and restore the land to the free farmer, of proposals to abolish debts in whole or in part. There was revolt and civil war. In 73 B.C., the distresses of Italy were enhanced by a great insurrection of the slaves under Spartacus. The slaves of Italy revolted with some effect, for among them were the trained fighters of the gladiatorial shows. For two years Spartacus held out in the crater of Vesuvius, which seemed at that time to be an extinct volcano. This insurrection was defeated at last and suppressed with frantic cruelty. Six thousand captured Spartacists were crucified along the Appian Way, the great highway that runs southward out of Rome (71 B.C.).

The common man never made head against the forces that were subjugating and degrading him. But the big rich men who were overcoming him were even in his defeat preparing a new power in the Roman world over themselves and him, the power of the army.

Before the Second Punic War the army of Rome was a levy of free farmers, who, according to their quality, rode or marched afoot to battle. This was a very good force for wars close at hand, but not the sort of army that will go abroad and bear long campaigns with patience. And moreover as the slaves multiplied and the estates grew, the supply of free-spirited fighting farmers declined. It was a popular leader named Marius who introduced a new factor. North Africa after the overthrow of the Carthaginian civilization had become a semi-barbaric kingdom, the kingdom of Numidia.

The Roman power fell into conflict with Jugurtha, king of this state, and experienced enormous difficulties in subduing him. Marius was made consul, in a phase of public indignation, to end this discreditable war. This he did by raising *paid troops* and drilling them hard. Jugurtha was brought in chains to Rome (106 B.C.) and Marius, when his time of office had expired, held on to his consulship illegally with his newly created legions. There was no power in Rome to restrain him.

With Marius began the third phase in the development of the Roman power, the Republic of the Military Commanders. For now began a period in which the leaders of the paid legions fought for the mastery of the Roman world. Against Marius was pitted the aristocratic Sulla who had served under him in Africa. Each in turn made a great massacre of his political opponents. Men were proscribed and executed by the thousand, and their estates were sold. After the bloody rivalry of these two and the horror of the revolt of Spartacus, came a phase in which Lucullus and Pompey the Great and Crassus and Julius Cæsar were the masters of armies and dominated affairs. It was Crassus who defeated Spartacus. Lucullus conquered Asia Minor and penetrated to Armenia, and retired with great wealth into private life. Crassus thrusting further invaded Persia and was defeated and slain by the Parthians. After a long rivalry Pompey was defeated by Julius Cæsar (48 B.C.) and murdered in Egypt, leaving Julius Cæsar sole master of the Roman world.

The figure of Julius Cæsar is one that has stirred the human imagination out of all proportion to its merit or true importance. He has become a legend and a symbol. For us he is chiefly important as marking the transition from the phase of military adventurers to the beginning of the fourth stage in Roman expansion, the Early Empire. For in spite of the profoundest economic and political convulsions, in spite of civil war and social degeneration, throughout all this time the boundaries of the Roman state crept outward and continued to creep outward to their maximum about 100 A.D. There had been something like an ebb during the doubtful phases of the Second Punic War, and again a manifest loss of vigour before the reconstruction of the army by Marius. The revolt of Spartacus marked a third phase. Julius Cæsar made his reputation as a military leader in Gaul, which is now France and Belgium. (The chief tribes inhabiting this country belonged to the same Celtic people as the Gauls who had occupied north Italy for a time, and who had afterwards raided into Asia Minor and settled down as the Galatians.) Cæsar drove back a German invasion of Gaul and added all that country to the empire, and he twice crossed the Straits of Dover into Britain (55 and 54 B.C.), where however he made no permanent conquest. Meanwhile Pompey the Great was consolidating Roman conquests that reached in the east to the Caspian Sea.

At this time, the middle of the first century B.C., the Roman Senate was still the nominal centre of the Roman government, appointing consuls and other officials, granting powers and the like; and a number of politicians, among whom Cicero was an outstanding

THE COLUMN OF TRAJAN AT ROME

Representing his conquests in Dacia and elsewhere

figure, were struggling to preserve the great traditions of republican Rome and to maintain respect for its laws. But the spirit of citizenship had gone from Italy with the wasting away of the free farmers; it was a land now of slaves and impoverished men with neither the understanding nor the desire for freedom. There was nothing whatever behind these republican leaders in the Senate, while behind the great adventurers they feared and desired to control were the legions. Over the heads of the Senate Crassus and Pompey and Cæsar divided the rule of the Empire between them (The First Triumvirate). When presently Crassus was killed at distant Carrhæ by the Parthians, Pompey and Cæsar fell out. Pompey took up the republican side, and laws were passed to bring Cæsar to trial for his breaches of law and his disobedience to the decrees of the Senate.

It was illegal for a general to bring his troops out of the boundary of his command, and the boundary between Cæsar's command and Italy was the Rubicon. In 49 B.C. he crossed the Rubicon, saying "The die is cast" and marched upon Pompey and Rome.

It had been the custom in Rome in the past, in periods of military extremity, to elect a "dictator" with practically unlimited powers to rule through the crisis. After his overthrow of Pompey, Cæsar was made dictator first for ten years and then (in 45 B.C.) for life. In effect he was made monarch of the empire for life. There was talk of a king, a word abhorrent to Rome since the expulsion of the Etruscans five centuries before. Cæsar refused to be king, but adopted throne and sceptre. After his defeat of Pompey, Cæsar had gone on into Egypt and had made love to Cleopatra, the last of the Ptolemies, the goddess queen of Egypt. She seems to have turned his head very completely. He had brought back to Rome the Egyptian idea of a god-king. His statue was set up in a temple with an inscription "To the Unconquerable God." The expiring republicanism of Rome flared up in a last protest, and Cæsar was stabbed to death in the Senate at the foot of the statue of his murdered rival, Pompey the Great.

Thirteen years more of this conflict of ambitious personalities followed. There was a second Triumvirate of Lepidus, Mark Antony and Octavian Cæsar, the latter the nephew of Julius Cæsar. Octavian like his uncle took the poorer, hardier western provinces where the best legions were recruited. In 31 B.C., he defeated Mark Antony, his only serious rival, at the naval battle of Actium, and made himself sole master of the Roman world. But Octavian was a man of different quality altogether from Julius Cæsar. He had no foolish craving to be God or King. He had no queen-lover that he wished to dazzle. He restored freedom to the Senate and people of Rome. He declined to be dictator. The grateful Senate in return gave him the reality instead of the forms of power. He was to be called not King indeed, but "Princeps" and "Augustus." He became Augustus Cæsar, the first of the Roman emperors (27 B.C. to 14 A.D.).

He was followed by Tiberius Cæsar (14 to 37 A.D.) and he by others, Caligula, Claudius, Nero and so on up to Trajan (98 A.D.), Hadrian (117 A.D.), Antonius Pius (138 A.D.) and Marcus Aurelius (161–180 A.D.). All these emperors were emperors of the legions. The soldiers made them, and some the soldiers destroyed. Gradually the Senate fades out of Roman history, and the emperor and his administrative officials replace it. The boundaries of the empire crept forward now to their utmost limits. Most of Britain was added to the empire, Transylvania was brought in as a new province, Dacia; Trajan crossed the Euphrates. Hadrian had an idea that reminds us at once of what had happened at the other end of the old world. Like Shi-Hwang-ti he built walls against the northern barbarians; one across Britain and a palisade between the Rhine and the Danube. He abandoned some of the acquisitions of Trajan.

The expansion of the Roman Empire was at an end.

XXXIV

Between Rome and China

THE second and first centuries B.C. mark a new phase in the history of mankind. Mesopotamia and the eastern Mediterranean are no longer the centre of interest. Both Mesopotamia and Egypt were still fertile, populous and fairly prosperous, but they were no longer the dominant regions of the world. Power had drifted to the west and to the east. Two great empires now dominated the world, this new Roman Empire and the renascent Empire of China. Rome extended its power to the Euphrates, but it was never able to get beyond that boundary. It was too remote. Beyond the Euphrates the former Persian and Indian dominions of the Seleucids fell under a number of new masters. China, now under the Han dynasty, which had replaced the Ts'in dynasty at the death of Shi-Hwang-ti, had extended its power across Tibet and over the high mountain passes of the Pamirs into western Turkestan. But there, too, it reached its extremes. Beyond was too far.

China at this time was the greatest, best organized and most civilized political system in the world. It was superior in area and population to the Roman Empire at its zenith. It was possible then for these two vast systems to flourish in the same world at the same time in almost complete ignorance of each other. The means of communication both by sea and land was not yet sufficiently developed and organized for them to come to a direct clash.

Yet they reacted upon each other in a very remarkable way, and their influence upon the fate of the regions that lay between them, upon central Asia and India, was profound. A certain amount of trade trickled through, by camel caravans across Persia, for example, and by coasting ships by way of India and the Red Sea. In 66 B.C. Roman troops under Pompey followed in the footsteps of Alexander the Great, and marched up the eastern shores of the

Caspian Sea. In 102 A.D. a Chinese expeditionary force under Pan Chau reached the Caspian, and sent emissaries to report upon the power of Rome. But many centuries were still to pass before definite knowledge and direct intercourse were to link the great parallel worlds of Europe and Eastern Asia.

To the north of both these great empires were barbaric wildernesses. What is now Germany was largely forest lands; the forests extended far into Russia and made a home for the gigantic aurochs, a bull of almost elephantine size. Then to the north of the great mountain masses of Asia stretched a band of deserts, steppes and then forests and frozen lands. In the eastward lap of the elevated part of Asia was the great triangle of Manchuria. Large parts of these regions, stretching between South Russia and Turkestan into Manchuria, were and are regions of exceptional climatic insecurity. Their rainfall has varied greatly in the course of a few centuries. They are lands treacherous to man. For years they will carry pasture and sustain cultivation, and then will come an age of decline in humidity and a cycle of killing droughts.

A CHINESE COVERED JAR OF GREEN-GLAZED EARTHENWARE

Han Dynasty (contemporary with late Roman republic and early Empire)

(In the Victoria and Albert Museum)

The western part of this barbaric north from the German forests to South Russia and Turkestan and from Gothland to the Alps was the region of origin of the Nordic peoples and of the Aryan speech. The eastern steppes and deserts of Mongolia was the region of origin of the Hunnish or Mongolian or Tartar or Turkish peoples — for all these several peoples were akin in language, race, and way of life. And as the Nordic peoples seem to have been continually overflowing their own borders and pressing south upon the developing civilizations of Mesopotamia and the Mediterranean coast, so the Hunnish tribes sent their surplus as wanderers, raiders and conquerors into the settled regions of China. Periods of plenty in the north would mean an increase in population there; a shortage of grass, a spell of cattle disease, would drive the hungry warlike tribesmen south.

For a time there were simultaneously two fairly effective Empires in the world capable of holding back the barbarians and even forcing forward the frontiers of the imperial peace. The thrust of the Han empire from north China into Mongolia was strong and continuous. The Chinese population welled up over the barrier of the Great Wall. Behind the imperial frontier guards came the Chinese farmer with horse and plough, ploughing up the grass lands and enclosing the winter pasture. The Hunnish peoples raided and murdered the settlers, but the Chinese punitive expeditions were too much for them. The nomads were faced with the choice of settling down to the plough and becoming Chinese tax-payers or shifting in search of fresh summer pastures. Some took the former course and were absorbed. Some drifted north-eastward and eastward over the mountain passes down into western Turkestan.

VASE OF BRONZE FORM, UNGLAZED STONEWARE
Han Dynasty (B.C. 206–A.D. 220)
(In the Victoria and Albert Museum)

This westward drive of the Mongolian horsemen was going on from 200 B.C. onward. It was producing a westward pressure upon the Aryan tribes, and these again were pressing upon the Roman frontiers ready to break through directly there was any weakness apparent. The Parthians, who were apparently a Scythian people with some Mongolian admixture, came down to the Euphrates by the first century B.C. They fought against Pompey the Great in his eastern raid. They defeated and killed Crassus. They replaced the Seleucid monarchy in Persia by a dynasty of Parthian kings, the Arsacid dynasty.

But for a time the line of least resistance for hungry nomads lay neither to the west nor the east but through central Asia and then south-eastward through the Khyber Pass into India. It was India which received the Mongolian drive in these centuries of Roman and Chinese strength. A series of raiding conquerors poured down through the Punjab into the great plains to loot and destroy. The

CHINESE VESSEL IN BRONZE, IN FORM OF A GOOSE

Dating from before the time of Shi-Hwang-ti. Such a piece of work indicates a high level of comfort and humour

(In the Victoria and Albert Museum)

empire of Asoka was broken up, and for a time the history of India passes into darkness. A certain Kushan dynasty founded by the "Indo-Scythians"—one of the raiding peoples—ruled for a time over North India and maintained a certain order. These invasions went on for several centuries. For a large part of the fifth century A.D. India was afflicted by the Ephthalites or White Huns, who levied tribute on the small Indian princes and held India in terror. Every summer these Ephthalites pastured in western Turkestan, every autumn they came down through the passes to terrorize India.

In the second century A.D. a great misfortune came upon the Roman and Chinese empires that probably weakened the resistance of both to barbarian pressure. This was a pestilence of unexampled virulence. It raged for eleven years in China and disorganized the social framework profoundly. The Han dynasty fell, and a new age of division and confusion began from which China did not fairly recover until the seventh century A.D. with the coming of the great Tang dynasty.

The infection spread through Asia to Europe. It raged throughout the Roman Empire from 164 to 180 A.D. It evidently weakened the Roman imperial fabric very seriously. We begin to hear of depopulation in the Roman provinces after this, and there was a marked deterioration in the vigour and efficiency of government. At any rate we presently find the frontier no longer invulnerable, but giving way first in this place and then in that. A new Nordic people, the Goths, coming originally from Gothland in Sweden, had migrated across Russia to the Volga region and the shores of the Black Sea and taken to the sea and piracy. By the end of the second century they may have begun to feel the westward thrust of the Huns. In 247 they crossed the Danube in a great land raid, and defeated and killed the Emperor Decius in a battle in what is now Serbia. In 236 another Germanic people, the Franks, had broken bounds upon the lower Rhine, and the Alemanni had poured into Alsace. The legions in Gaul beat back their invaders, but the Goths in the Balkan peninsula raided again and again. The province of Dacia vanished from Roman history.

A chill had come to the pride and confidence of Rome. In 270–275 Rome, which had been an open and secure city for three centuries, was fortified by the Emperor Aurelian.

XXXV

The Common Man's Life under the Early Roman Empire

BEFORE we tell of how this Roman empire which was built up in the two centuries B.C., and which flourished in peace and security from the days of Augustus Cæsar onward for two centuries, fell into disorder and was broken up, it may be as well to devote some attention to the life of the ordinary people throughout this great realm. Our history has come down now to within 2000 years of our own time; and the life of the civilized people, both under the Peace of Rome and the Peace of the Han dynasty, was beginning to resemble more and more clearly the life of their civilized successors to-day.

In the western world coined money was now in common use; outside the priestly world there were many people of independent means who were neither officials of the government nor priests; people travelled about more freely than they had ever done before, and there were high roads and inns for them. Compared with the past, with the time before 500 B.C., life had become much more loose. Before that date civilized men had been bound to a district or country, had been bound to a tradition and lived within a very limited horizon; only the nomads traded and travelled.

But neither the Roman Peace nor the Peace of the Han dynasty meant a uniform civilization over the large areas they controlled. There were very great local differences and great contrasts and inequalities of culture between one district and another, just as there are to-day under the British Peace in India. The Roman garrisons and colonies were dotted here and there over this great space, worshipping Roman gods and speaking the Latin language; but where there had been towns and cities before the coming of the Romans, they went on, subordinated indeed but managing their own affairs, and, for a time at least, worshipping their own gods in their own fashion. Over Greece, Asia Minor, Egypt and the Hellenized East generally, the Latin language never prevailed. Greek ruled there invincibly. Saul of Tarsus, who became the apostle Paul, was a Jew and a Roman citizen; but he spoke and wrote Greek and not Hebrew. Even at the court of the Parthian dynasty, which had overthrown the Greek Seleucids in Persia, and was quite outside the Roman imperial boundaries, Greek was the fashionable language.

In some parts of Spain and in North Africa, the Carthaginian language also held on for a long time in spite of the destruction of Carthage. Such a town as Seville, which had been a prosperous city long before the Roman name had been heard of, kept its Semitic goddess and preserved its Semitic speech for generations, in spite of a colony of Roman veterans at Italica a few miles away. Septimius Severus, who was emperor from 193 to 211 A.D., spoke Carthaginian as his mother speech. He learnt Latin later as a foreign tongue;

and it is recorded that his sister never learnt Latin and conducted her Roman household in the Punic language.

In such countries as Gaul and Britain and in provinces like Dacia (now roughly Roumania) and Pannonia (Hungary south of the Danube), where there were no pre-existing great cities and temples and cultures, the Roman empire did however "Latinize." It civilized these countries for the first time. It created cities and towns where Latin was from the first the dominant speech, and where Roman gods were served and Roman customs and fashions followed. The Roumanian, Italian, French and Spanish languages, all variations and modifications of Latin, remain to remind us of this extension of Latin speech and customs. North-west Africa also became at last largely Latin-speaking. Egypt, Greece and the rest of the empire to the east were never Latinized. They remained Egyptian and Greek in culture and spirit. And even in Rome, among educated men, Greek was learnt as the language of a gentleman and Greek literature and learning were very properly preferred to Latin.

In this miscellaneous empire the ways of doing work and business were naturally also very miscellaneous. The chief industry of the settled world was still largely agriculture. We have told how in Italy the sturdy free farmers who were the backbone of the early Roman republic were replaced by estates worked by slave labour after the Punic wars. The Greek world had had very various methods of cultivation, from the Arcadian plan, wherein every free citizen toiled with his own hands, to Sparta, wherein it was a dishonour to work and where agricultural work was done by a special slave class, the Helots. But that was ancient history now, and over most of the Hellenized world the estate system and slave-gangs had spread. The agricultural slaves were captives who spoke many different languages so that they could not understand each other, or they were born slaves; they had no solidarity to resist oppression, no tradition of rights, no knowledge, for they could not read nor write. Although they came to form a majority of the country population they never made a successful insurrection. The insurrection of Spartacus in the first century B.C. was an insurrection of the special slaves who were trained for the gladiatorial combats. The agricultural workers in Italy in the latter days of the Republic and the early Empire suffered frightful indignities; they would be chained at night to prevent escape or have half the head shaved to make it difficult. They had no wives of their own; they could be outraged, mutilated and killed by their masters. A master could sell his slave to fight beasts in the arena. If a slave slew his master, all the slaves in his household and not merely the murderer were crucified. In some parts of Greece, in Athens notably, the lot of the slave was never quite so frightful as this, but it was still detestable. To such a population the barbarian invaders who presently broke through the defensive line of the legions, came not as enemies but as liberators.

POMPEII

"Note the ruts in roadway worn by chariot wheels."

The slave system had spread to most industries and to every sort of work that could be done by gangs. Mines and metallurgical operations, the rowing of galleys, road-making and big building operations were all largely slave occupations. And almost all domestic service was performed by slaves. There were poor freemen and there were freed-men in the cities and upon the country side, working for themselves or even working for wages. They were artizans, supervisors and so forth, workers of a new money-paid class working in competition with slave workers; but we do not know what proportion they made of the general population. It probably varied widely in different places and at different periods. And there were also many modifications of slavery, from the slavery that was chained at night and driven with whips to the farm or quarry, to the slave whose master found it advantageous to leave him to cultivate his patch or work his craft and own his wife like a free-man, provided he paid in a satisfactory quittance to his owner.

There were armed slaves. At the opening of the period of the Punic wars, in 264 B.C., the Etruscan sport of setting slaves to fight for their lives was revived in Rome. It grew rapidly fashionable; and soon every great Roman rich man kept a retinue of gladiators, who sometimes fought in the arena but whose real business it was to act as his bodyguard of bullies. And also there were learned slaves. The conquests of the later Republic were among the highly civilized cities of Greece, North Africa and Asia Minor; and they brought in many highly educated captives. The tutor of a young Roman of good family was usually a slave. A rich man would have a Greek slave as librarian, and slave secretaries and learned men. He would keep his poet as he would keep a performing dog. In this atmosphere of slavery the traditions of modern literary criticism were evolved. The slaves still boast and quarrel in our reviews. There were enterprising people who bought intelligent boy slaves and had them educated for sale. Slaves were trained as book copyists, as jewellers, and for endless skilled callings.

But there were very considerable changes in the position of a slave during the four hundred years between the opening days of conquest under the republic of rich men and the days of disintegration that followed the great pestilence. In the second century B.C. war-captives were abundant, manners gross and brutal; the slave had no rights and there was scarcely an outrage the reader can imagine that was not practised upon slaves in those days. But already in the first century A.D. there was a perceptible improvement in the attitude of the Roman civilization towards slavery. Cap-

Photo: Underwood & Underwood

THE COLISEUM, ROME

INTERIOR OF THE COLISEUM AS IT APPEARS TO-DAY

tives were not so abundant for one thing, and slaves were dearer. And slave-owners began to realize that the profit and comfort they got from their slaves increased with the self-respect of these unfortunates. But also the moral tone of the community was rising, and a sense of justice was becoming effective. The higher mentality of Greece was qualifying the old Roman harshness. Restrictions upon cruelty were made, a master might no longer sell his slave to fight beasts, a slave was given property rights in what was called his *peculium*, slaves were paid wages as an encouragement and stimulus, a form of slave marriage was recognized. Very many forms of agriculture do not lend themselves to gang working, or require gang workers only at certain seasons. In regions where such conditions prevailed the slave presently became a serf, paying his owner part of his produce or working for him at certain seasons.

When we begin to realize how essentially this great Latin and Greek-speaking Roman Empire of the first two centuries A.D. was a slave state and how small was the minority who had any pride or freedom in their lives, we lay our hands on the clues to its decay and collapse. There was little of what we should call family life, few homes of temperate living and active thought and study; schools and colleges were few and far between. The free will and the free mind were nowhere to be found. The great roads, the ruins of splendid buildings, the tradition of law and power it left for the astonishment of succeeding generations must not conceal from us that all its outer splendour was built upon thwarted wills, stifled intelligence, and crippled and perverted desires. And even the minority who lorded it over that wide realm of subjugation and of restraint and forced labour were uneasy and unhappy in their souls; art and literature, science and philosophy, which are the fruits of free and happy minds, waned in that atmosphere. There was much copying and imitation, an abundance of artistic artificers, much slavish pedantry among the servile men of learning, but the whole Roman empire in four centuries produced nothing to set beside the bold and noble intellectual activities of the comparatively little city of Athens during its one century of greatness. Athens decayed under the Roman sceptre. The science of Alexandria decayed. The spirit of man, it seemed, was decaying in those days.

XXXVI

Religious Developments under the Roman Empire

THE soul of man under that Latin and Greek empire of the first two centuries of the Christian era was a worried and frustrated soul. Compulsion and cruelty reigned; there were pride and display but little honour; little serenity or steadfast happiness. The unfortunate were despised and wretched; the fortunate were insecure and feverishly eager for gratifications. In a great number of cities life centred on the red excitement of the arena, where men and beasts fought and were tormented and slain. Amphitheatres are the most characteristic of Roman ruins. Life went on in that key. The uneasiness of men's hearts manifested itself in profound religious unrest.

From the days when the Aryan hordes first broke in upon the ancient civilizations, it was inevitable that the old gods of the temples and priesthoods should suffer great adaptations or disappear. In the course of hundreds of generations the agricultural peoples of the brunette civilizations had shaped their lives and thoughts to the temple-centred life. Observances and the fear of disturbed routines, sacrifices and mysteries, dominated their minds. Their gods seem monstrous and illogical to our modern minds because we belong to an Aryanized world, but to these older peoples these deities had the immediate conviction and vividness of things seen in an intense dream. The conquest of one city state by another in Sumeria or early Egypt meant a change or a renaming of gods or goddesses, but left the shape and spirit of the worship intact. There was no change in its general character. The figures in the dream changed, but the dream went on and it was the same sort of dream. And the early Semitic conquerors were sufficiently akin in spirit to the Sumerians to take over the religion of the Mesopotamian civilization they subjugated without any profound alteration. Egypt was never indeed subjugated to the extent of a religious revolution. Under the Ptolemies and under the Cæsars, her temples and altars and priesthoods remained essentially Egyptian.

So long as conquests went on between people of similar social and religious habits it was possible to get over the clash between the god of this temple and region and the god of that by a process of grouping or assimilation. If the two gods were alike in character they were identified. It was really the same god under another name, said the priests and the people. This fusion of gods is called theocrasia; and the age of the great conquests of the thousand years B.C. was an age of theocrasia. Over wide areas the local gods were displaced by, or rather they were swallowed up in, a general god. So that when at last Hebrew prophets in Babylon proclaimed one God of Righteousness in all the earth men's minds were fully prepared for that idea.

But often the gods were too dissimilar for such an assimilation, and then they were grouped together in some plausible relationship. A female god — and the Ægean world before the coming of the Greek was much addicted to Mother Gods — would be married to a male god, and an animal god or a star god would be humanized and the animal or astronomical aspect, the serpent or the sun or the star, made into an ornament or a symbol. Or the god of a defeated people would become a malignant antagonist to the brighter gods. The history of theology is full of such adaptations, compromises and rationalizations of once local gods.

As Egypt developed from city states into one united kingdom there was much of this theocrasia. The chief god so to speak was Osiris, a sacrificial harvest god of whom Pharaoh was supposed to be the earthly incarnation. Osiris was represented as repeatedly dying and rising again; he was not only the seed and the harvest but also by a natural extension of thought the means of human immortality. Among his symbols was the wide-winged scarabeus beetle which buries its eggs to rise again, and also the effulgent sun which sets to rise. Later on he was to be identified with Apis, the sacred bull. Associated with him was the goddess Isis. Isis was also Hathor, a cow-goddess, and the crescent moon and the Star of the sea. Osiris dies and she bears a child, Horus, who is also a hawk-god and the dawn, and who grows to become Osiris again. The effigies of Isis represent her as bearing the infant Horus in her arms and standing on the crescent moon. These are not logical relationships, but they were devised by the human mind before the development of hard and systematic thinking and they have a dream-like coherence. Beneath this triple group there are other and darker Egyptian gods, bad gods, the dog-headed Anubis, black night and the like, devourers, tempters, enemies of god and man.

Every religious system does in the course of time fit itself to the shape of the human soul, and there can be no doubt that out of these illogical and even uncouth symbols, Egyptian people were able to fashion for themselves ways of genuine devotion and consolation. The desire for immortality was very strong in the Egyptian mind, and the religious life of Egypt turned on that desire. The Egyp-

MITHRAS SACRIFICING A BULL, ROMAN

(In the British Museum)

tian religion was an immortality religion as no other religion had ever been. As Egypt went down under foreign conquerors and the Egyptian gods ceased to have any satisfactory political significance, this craving for a life of compensations hereafter, intensified.

ISIS AND HORUS

After the Greek conquest, the new city of Alexandria became the centre of Egyptian religious life, and indeed of the religious life of the whole Hellenic world. A great temple, the Serapeum, was set up by Ptolemy I at which a sort of trinity of gods was worshipped. These were Serapis (who was Osiris-Apis rechristened), Isis and Horus. These were not regarded as separate gods but as three aspects of one god, and Serapis was identified with the Greek Zeus, the Roman Jupiter and the Persian sun-god. This worship spread wherever the Hellenic influence extended, even into North India and Western China. The idea of immortality, an immortality of compensations and consolation, was eagerly received by a world in which the common life was hopelessly wretched. Serapis was called "the saviour of souls." "After death," said the hymns of that time, "we are still in the care of his providence." Isis attracted many devotees. Her images stood in her temples, as Queen of Heaven, bearing the infant Horus in her arms. Candles were burnt before her, votive offerings were made to her, shaven priests consecrated to celibacy waited on her altar.

The rise of the Roman empire opened the western European world to this growing cult. The temples of Serapis-Isis, the chanting of the priests and the hope of immortal life, followed the Roman standards to Scotland and Holland. But there were many rivals to the Serapis-Isis religion. Prominent among these was Mithraism. This was a religion of Persian origin, and it centred upon some now forgotten mysteries about Mithras sacrificing a sacred and benevolent bull. Here we seem to have something more primordial than the complicated and sophisticated Serapis-Isis beliefs. We are carried back directly to the blood sacrifices of the heliolithic stage in human culture. The bull upon the Mithraic monuments always bleeds copiously from a wound in its side, and from this blood springs new life. The votary to Mithraism actually bathed in the blood of the sacrificial bull. At his initiation he went beneath a scaffolding upon which a bull was killed so that the blood could actually run down on him.

Both these religions, and the same is true of many other of the numerous parallel cults that sought the allegiance of the slaves and citizens under the earlier Roman emperors, are personal religions. They aim at personal salvation and personal immortality. The older religions were not personal like that; they were social. The older fashion of divinity was god or goddess of the city first or of the state, and only secondarily of the individual. The sacrifices were a public and not a private function. They concerned collective practical needs in this world in which we live. But the Greeks first and now the Romans had pushed religion out of politics. Guided by the Egyptian tradition religion had retreated to the other world.

These new private immortality religions took all the heart and emotion out of the old state religions, but they did not actually replace them. A typical city under the earlier Roman emperors would have a number of temples to all sorts of gods. There might be a temple to Jupiter of the Capitol, the great god of Rome, and there would probably be one to the reigning Cæsar. For the Cæsars had learnt from the Pharaohs the possibility of being gods. In such temples a cold and stately political worship went on; one would go and make an offering and burn a pinch of incense to show one's loyalty. But it would be to the temple of Isis, the dear Queen of Heaven, one would go with the burthen

BUST OF THE EMPEROR COMMODUS, A.D. 180–192

Represented as the God Mithras, Roman, Circa A.D. 190

(In the British Museum)

of one's private troubles for advice and relief. There might be local and eccentric gods. Seville, for example, long affected the worship of the old Carthaginian Venus. In a cave or an underground temple there would certainly be an altar to Mithras, attended by legionaries and slaves. And probably also there would be a synagogue where the Jews gathered to read their Bible and uphold their faith in the unseen God of all the Earth.

Sometimes there would be trouble with the Jews about the political side of the state religion. They held that their God was a jealous God intolerant of idolatry, and they would refuse to take part in the public sacrifices to Cæsar. They would not even salute the Roman standards for fear of idolatry.

In the East long before the time of Buddha there had been ascetics, men and women who gave up most of the delights of life, who repudiated marriage and property and sought spiritual powers and an escape from the stresses and mortifications of the world in abstinence, pain and solitude. Buddha himself set his face against ascetic extravagances, but many of his disciples followed a monkish life of great severity. Obscure Greek cults practised similar disciplines even to the extent of self-mutilation. Asceticism appeared in the Jewish communities of Judea and Alexandria also in the first century B.C. Communities of men abandoned the world and gave themselves to austerities and mystical contemplation. Such was the sect of the Essenes. Throughout the first and second centuries A.D. there was an almost world-wide resort to such repudiations of life, a universal search for "salvation" from the distresses of the time. The old sense of an established order, the old confidence in priest and temple and law and custom, had gone. Amidst the prevailing slavery, cruelty, fear, anxiety, waste, display and hectic self-indulgence, went this epidemic of self-disgust and mental insecurity, this agonized search for peace even at the price of renunciation and voluntary suffering. This it was that filled the Serapeum with weeping penitents and brought the converts into the gloom and gore of the Mithraic cave.

XXXVII

THE TEACHING OF JESUS

IT was while Augustus Cæsar, the first of the Emperors, was reigning in Rome that Jesus who is the Christ of Christianity was born in Judea. In his name a religion was to arise which was destined to become the official religion of the entire Roman Empire.

Now it is on the whole more convenient to keep history and theology apart. A large proportion of the Christian world believes that Jesus was an incarnation of that God of all the Earth whom the Jews first recognized. The historian, if he is to remain historian, can neither accept nor deny that interpretation. Materially Jesus appeared in the likeness of a man, and it is as a man that the historian must deal with him.

He appeared in Judea in the reign of Tiberius Cæsar. He was a prophet. He preached after the fashion of the preceding Jewish prophets. He was a man of about thirty, and we are in the profoundest ignorance of his manner of life before his preaching began.

Our only direct sources of information about the life and teaching of Jesus are the four Gospels. All four agree in giving us a picture of a very definite personality. One is obliged to say, "Here was a man. This could not have been invented."

But just as the personality of Gautama Buddha has been distorted and obscured by the stiff squatting figure, the gilded idol of later Buddhism, so one feels that the lean and strenuous personality of Jesus is much wronged by the unreality and conventionality that a mistaken reverence has imposed upon his figure in modern Christian art. Jesus was a penniless teacher, who wandered about the dusty sun-bit country of Judea, living upon casual gifts of food; yet he is always represented clean, combed and sleek, in spotless raiment, erect and with something motionless about him as though he was gliding through the air. This alone has made him unreal and incredible to many people who cannot distinguish the core of the story from the ornamental and unwise additions of the unintelligently devout.

We are left, if we do strip this record of these difficult accessories, with the figure of a being, very human, very earnest and passionate, capable of swift anger, and teaching a new and simple and profound doctrine — namely, the universal loving Fatherhood of God and the coming of the Kingdom of Heaven. He was clearly a person — to use a common phrase — of intense personal magnetism. He attracted followers and filled them with love and courage. Weak and ailing people were heartened and healed by his presence. Yet he was probably of a delicate physique, because of the swiftness with which he died under the pains of crucifixion. There is a tradition that he fainted when, according to the custom, he was made to bear his cross to the place of execution. He went about the country for three years spreading his doctrine and then he came to Jerusalem and was accused of trying to set up a strange kingdom in Judea; he was tried upon this charge, and crucified together with two thieves. Long before these two were dead his sufferings were over.

The doctrine of the Kingdom of Heaven, which was the main teaching of Jesus, is certainly one of the most revolutionary doctrines that ever stirred and changed human thought. It is small wonder if the world of that time failed to grasp its full significance, and recoiled in dismay from even a half apprehension of its tremendous challenges to the established habits and institutions of mankind. For the doctrine of the Kingdom of Heaven, as Jesus seems to have preached it, was no less than a bold and uncompromising demand for a complete change and cleansing of the life of our struggling race, an utter cleansing, without and within. To the gospels the reader must go for all that is preserved of this tremendous teaching; here we are only concerned with the jar of its impact upon established ideas.

The Jews were persuaded that God, the one God of the whole world, was a righteous god, but they also thought of him as a trading god who had made a bargain with their Father Abraham

EARLY IDEAL PORTRAIT, IN GILDED GLASS, OF JESUS CHRIST IN WHICH THE TRADITIONAL BEARD IS NOT SHOWN

about them, a very good bargain indeed for them, to bring them at last to predominance in the earth. With dismay and anger they heard Jesus sweeping away their dear securities. God, he taught, was no bargainer; there were no chosen people and no favourites in the Kingdom of Heaven. God was the loving father of all life, as incapable of showing favour as the universal sun. And all men were brothers — sinners alike and beloved sons alike — of this divine father. In the parable of the Good Samaritan Jesus cast scorn upon that natural tendency we all obey, to glorify our own people and to minimize the righteousness of other creeds and other races. In the parable of the labourers he thrust aside the obstinate claim of the Jews to have a special claim upon God. All whom God takes into the kingdom, he taught, God serves alike; there is no distinction in his treatment, because there is no measure to his bounty. From all, moreover, as the parable of the buried talent witnesses, and as the incident of the widow's mite enforces, he demands the utmost. There are no privileges, no rebates and no excuses in the Kingdom of Heaven.

But it is not only the intense tribal patriotism of the Jews that Jesus outraged. They were a people of intense family loyalty, and he would have swept away all the narrow and restrictive family affections in the great flood of the love of God. The whole kingdom of Heaven was to be the family of his followers. We are told that, "While he yet talked to the people, behold, his mother and his brethren stood without, desiring to speak with him. Then one said unto him, Behold, thy mother and thy brethren stand without, desiring to speak with thee. But he answered and said unto him that told him, Who is my mother? and who are my brethren? And he stretched forth his hands towards his disciples, and said, Behold my mother and my brethren! For whosoever shall do the will of my Father which is in heaven, the same is my brother, and sister, and mother."[1]

And not only did Jesus strike at patriotism and the bonds of family loyalty in the name of God's universal fatherhood and brotherhood of all mankind, but it is clear that his teaching condemned all the gradations of the economic system, all private wealth, and

[1] Matt. xii. 46–50.

Photo: *Jannaway*

THE ROAD FROM NAZARETH TO TIBERIAS

personal advantages. All men belonged to the kingdom; all their possessions belonged to the kingdom; the righteous life for all men, the only righteous life, was the service of God's will with all that we had, with all that we were. Again and again he denounced private riches and the reservation of any private life.

"And when he was gone forth into the way, there came one running, and kneeled to him, and asked him, Good Master, what shall I do that I may inherit eternal life? And Jesus said to him, Why callest thou me good? there is none good but one, that is God. Thou knowest the commandments, Do not commit adultery, Do not kill, Do not steal, Do not bear false witness, Defraud not, Honour thy father and mother. And he answered and said unto him, Master, all these things have I observed from my youth. Then Jesus beholding him loved him, and said unto him, One thing thou lackest; go thy way, sell whatsoever thou hast, and give to the poor, and thou shalt have treasure in heaven: and come, take up the cross, and follow me. And he was sad at that saying, and went away grieved: for he had great possessions.

Photo: Jannaway

DAVID'S TOWER AND WALL OF JERUSALEM

"And Jesus looked round about, and saith unto his disciples, How hardly shall they that have riches enter into the Kingdom of God! And the disciples were astonished at his words. But Jesus answered again, and saith unto them, Children, how hard is it for them that trust in riches to enter into the Kingdom of God! It is

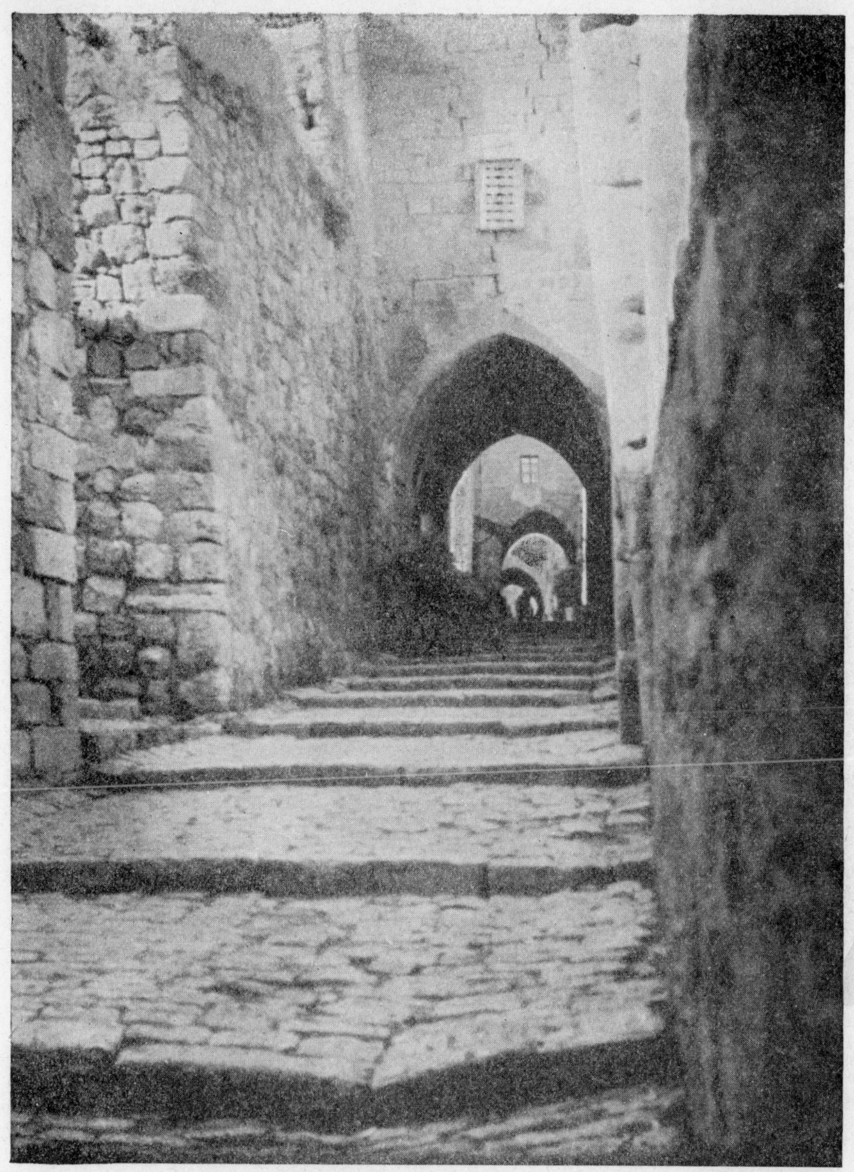

Photo: *Jannaway*

A STREET IN JERUSALEM

Along such a thoroughfare Christ carried his cross to the place of execution

easier for a camel to go through the eye of a needle, than for a rich man to enter into the Kingdom of God." [1]

Moreover, in his tremendous prophecy of this kingdom which was to make all men one together in God, Jesus had small patience for the bargaining righteousness of formal religion. Another large part of his recorded utterances is aimed against the meticulous observance of the rules of the pious career. "Then the Pharisees and scribes asked him, Why walk not thy disciples according to the tradition of the elders, but eat bread with unwashen hands? He answered and said unto them, Well hath Isaiah prophesied of you hypocrites, as it is written,

"This people honoureth me with their lips,
"But their heart is far from me.
"Howbeit in vain do they worship me,
"Teaching for doctrines the commandments of men.

"For laying aside the commandment of God, ye hold the tradition of men, as the washing of pots and cups: and many other such things ye do. And he said unto them, Full well ye reject the commandment of God, that ye may keep your own tradition." [2]

It was not merely a moral and a social revolution that Jesus proclaimed; it is clear from a score of indications that his teaching had a political bent of the plainest sort. It is true that he said his kingdom was not of this world, that it was in the hearts of men and not upon a throne; but it is equally clear that wherever and in what measure his kingdom was set up in the hearts of men, the outer world would be in that measure revolutionized and made new.

Whatever else the deafness and blindness of his hearers may have missed in his utterances, it is plain they did not miss his resolve to revolutionize the world. The whole tenor of the opposition to him and the circumstances of his trial and execution show clearly that to his contemporaries he seemed to propose plainly, and did propose plainly, to change and fuse and enlarge all human life.

In view of what he plainly said, is it any wonder that all who were rich and prosperous felt a horror of strange things, a swimming of their world at his teaching? He was dragging out all the little private reservations they had made from social service into the light

[1] Mark x. 17–25. [2] Mark vii. 1–9.

of a universal religious life. He was like some terrible moral huntsman digging mankind out of the snug burrows in which they had lived hitherto. In the white blaze of this kingdom of his there was to be no property, no privilege, no pride and precedence; no motive indeed and no reward but love. Is it any wonder that men were dazzled and blinded and cried out against him? Even his disciples cried out when he would not spare them the light. Is it any wonder that the priests realized that between this man and themselves there was no choice but that he or priestcraft should perish? Is it any wonder that the Roman soldiers, confronted and amazed by something soaring over their comprehension and threatening all their disciplines, should take refuge in wild laughter, and crown him with thorns and robe him in purple and make a mock Cæsar of him? For to take him seriously was to enter upon a strange and alarming life, to abandon habits, to control instincts and impulses, to essay an incredible happiness. . . .

XXXVIII

The Development of Doctrinal Christianity

IN the four gospels we find the personality and teachings of Jesus but very little of the dogmas of the Christian church. It is in the epistles, a series of writings by the immediate followers of Jesus, that the broad lines of Christian belief are laid down.

Chief among the makers of Christian doctrine was St. Paul. He had never seen Jesus nor heard him preach. Paul's name was originally Saul, and he was conspicuous at first as an active persecutor of the little band of disciples after the crucifixion. Then he was suddenly converted to Christianity, and he changed his name to Paul. He was a man of great intellectual vigour and deeply and passionately interested in the religious movements of the time. He was well versed in Judaism and in the Mithraism and Alexandrian religion of the day. He carried over many of their ideas and terms of expression into Christianity. He did very little to enlarge or develop the original teaching of Jesus, the teaching of the Kingdom of Heaven. But he taught that Jesus was not only the promised Christ, the promised leader of the Jews, but also that his death was a sacrifice, like the deaths of the ancient sacrificial victims of the primordial civilizations, for the redemption of mankind.

When religions flourish side by side they tend to pick up each other's ceremonial and other outward peculiarities. Buddhism, for example, in China has now almost the same sort of temples and priests and uses as Taoism, which follows in the teachings of Lao Tse. Yet the original teachings of Buddhism and Taoism were almost flatly opposed. And it reflects no doubt or discredit upon the essentials of Christian teaching that it took over not merely such formal things as the shaven priest, the votive offering, the altars, candles, chanting and images of the Alexandrian and Mithraic faiths, but adopted even their devotional phrases and their theo-

MOSAIC OF SS. PETER AND PAUL POINTING TO A THRONE, ON GOLD BACKGROUND
From the Ninth Century original, in the Church of Sta. Prassede, Rome
(*In the Victoria and Albert Museum*)

logical ideas. All these religions were flourishing side by side with many less prominent cults. Each was seeking adherents, and there must have been a constant going and coming of converts between them. Sometimes one or other would be in favour with the government. But Christianity was regarded with more suspicion than its rivals because, like the Jews, its adherents would not perform acts of worship to the God Cæsar. This made it a seditious religion, quite apart from the revolutionary spirit of the teachings of Jesus himself.

St. Paul familiarized his disciples with the idea that Jesus, like

Osiris, was a god who died to rise again and give men immortality. And presently the spreading Christian community was greatly torn by complicated theological disputes about the relationship of this God Jesus to God the Father of Mankind. The Arians taught that Jesus was divine, but distant from and inferior to the Father. The Sabellians taught that Jesus was merely an aspect of the Father, and that God was Jesus and Father at the same time just as a man may be a father and an artificer at the same time; and the Trinitarians taught a more subtle doctrine that God was both one and three, Father, Son and Holy Spirit. For a time it seemed that Arianism would prevail over its rivals, and then after disputes, violence and wars, the Trinitarian formula became the accepted formula of all Christendom. It may be found in its completest expression in the Athanasian Creed.

We offer no comment on these controversies here. They do not sway history as the personal teaching of Jesus sways history. The personal teaching of Jesus does seem to mark a new phase in the moral and spiritual life of our race. Its insistence upon the universal Fatherhood of God and the implicit brotherhood of all men, its insistence upon the sacredness of every human personality as a living temple of God, was to have the profoundest effect upon all the subsequent social and political life of mankind. With Christianity, with the spreading teachings of Jesus, a new respect appears in the world for man as man. It may be true, as hostile critics of Christianity have urged, that St. Paul preached obedience to slaves, but it is equally true that the whole spirit of the teachings of Jesus preserved in the gospels was against the subjugation of man by man. And still more distinctly was Christianity opposed to such outrages upon human dignity as the gladiatorial combats in the arena.

Throughout the first two centuries after Christ, the Christian religion spread throughout the Roman Empire, weaving together an ever-growing multitude of converts into a new community of ideas and will. The attitude of the emperors varied between hostility and toleration. There were attempts to suppress this new faith in both the second and third centuries; and finally in 303 and the following years a great persecution under the Emperor Diocletian. The considerable accumulations of Church property were

THE BAPTISM OF CHRIST
(Sixth Century Ivory Panel in the British Museum)

seized, all bibles and religious writings were confiscated and destroyed, Christians were put out of the protection of the law and many executed. The destruction of the books is particularly notable. It shows how the power of the written word in holding together the new faith was appreciated by the authorities. These "book religions," Christianity and Judaism, were religions that educated. Their continued existence depended very largely on people being able to read and understand their doctrinal ideas. The older religions had made no such appeal to the personal intelligence. In the ages of barbaric confusion that were now at hand in western Europe it was the Christian church that was mainly instrumental in preserving the tradition of learning.

The persecution of Diocletian failed completely to suppress the growing Christian community. In many provinces it was ineffective because the bulk of the population and many of the officials were Christian. In 317 an edict of toleration was issued by the associated Emperor Galerius, and in 324 Constantine the Great, a friend and on his deathbed a baptized convert to Christianity, became sole ruler of the Roman world. He abandoned all divine pretensions and put Christian symbols on the shields and banners of his troops.

In a few years Christianity was securely established as the official religion of the empire. The competing religions disappeared or were absorbed with extraordinary celerity, and in 390 Theodosius the Great caused the great statue of Jupiter Serapis at Alexandria to be destroyed. From the outset of the fifth century onward the only priests or temples in the Roman Empire were Christian priests and temples.

XXXIX

The Barbarians Break the Empire into East and West

THROUGHOUT the third century the Roman Empire, decaying socially and disintegrating morally, faced the barbarians. The emperors of this period were fighting military autocrats, and the capital of the empire shifted with the necessities of their military policy. Now the imperial headquarters would be at Milan in north Italy, now in what is now Serbia at Sirmium or Nish, now in Nicomedia in Asia Minor. Rome halfway down Italy was too far from the centre of interest to be a convenient imperial seat. It was a declining city. Over most of the empire peace still prevailed and men went about without arms. The armies continued to be the sole repositories of power; the emperors, dependent on their legions, became more and more autocratic to the rest of the empire and their state more and more like that of the Persian and other oriental monarchs. Diocletian assumed a royal diadem and oriental robes.

All along the imperial frontier, which ran roughly along the Rhine and Danube, enemies were now pressing. The Franks and other German tribes had come up to the Rhine. In north Hungary were the Vandals; in what was once Dacia and is now Roumania, the Visigoths or West Goths. Behind these in south Russia were the East Goths or Ostrogoths, and beyond these again in the Volga region the Alans. But now Mongolian peoples were forcing their way towards Europe. The Huns were already exacting tribute from the Alans and Ostrogoths and pushing them to the west.

In Asia the Roman frontiers were crumpling back under the push of a renascent Persia. This new Persia, the Persia of the Sassenid kings, was to be a vigorous and on the whole a successful rival of the Roman Empire in Asia for the next three centuries.

A glance at the map of Europe will show the reader the peculiar weakness of the empire. The river Danube comes down to within a couple of hundred miles of the Adriatic Sea in the region of what is now Bosnia and Serbia. It makes a square re-entrant angle there. The Romans never kept their sea communications in good order, and this two hundred mile strip of land was their line of communication between the western Latin-speaking part of the empire and the eastern Greek-speaking portion. Against this square angle of the Danube the barbarian pressure was greatest. When they broke through there it was inevitable that the empire should fall into two parts.

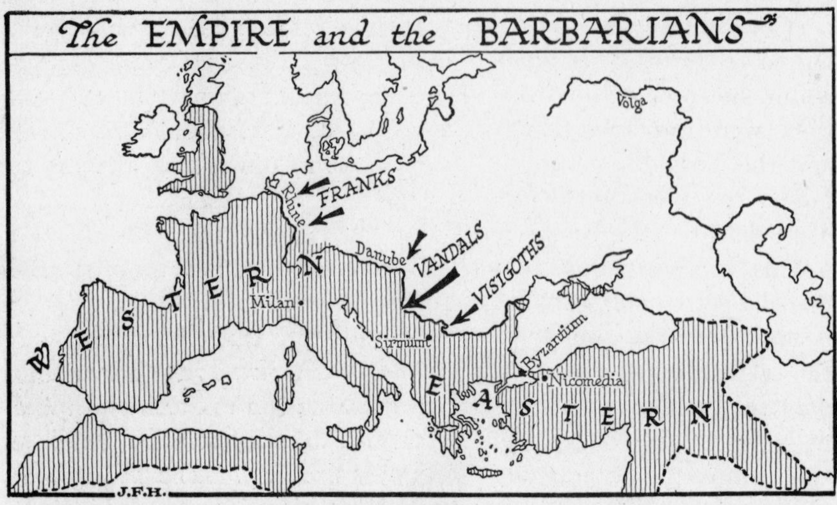

A more vigorous empire might have thrust forward and reconquered Dacia, but the Roman Empire lacked any such vigour. Constantine the Great was certainly a monarch of great devotion and intelligence. He beat back a raid of the Goths from just these vital Balkan regions, but he had no force to carry the frontier across the Danube. He was too pre-occupied with the internal weaknesses of the empire. He brought the solidarity and moral force of Christianity to revive the spirit of the declining empire, and he decided to create a new permanent capital at Byzantium upon the Hellespont. This new-made Byzantium, which was re-christened Constantinople in his honour, was still building when he died. Towards the end of his reign occurred a remarkable transaction. The

Vandals, being pressed by the Goths, asked to be received into the Roman Empire. They were assigned lands in Pannonia, which is now that part of Hungary west of the Danube, and their fighting men became nominally legionaries. But these new legionaries remained under their own chiefs. Rome failed to digest them.

Photo: Sebah & Joaillier

CONSTANTINE'S PILLAR, CONSTANTINOPLE

Constantine died working to reorganize his great realm, and soon the frontiers were ruptured again and the Visigoths came almost to Constantinople. They defeated the Emperor Valens at Adrianople and made a settlement in what is now Bulgaria, similar to the settlement of the Vandals in Pannonia. Nominally they were subjects of the emperor, practically they were conquerors.

From 379 to 395 A.D. reigned the Emperor Theodosius the Great, and while he reigned the empire was still formally intact. Over the armies of Italy and Pannonia presided Stilicho, a Vandal, over the armies in the Balkan peninsula, Alaric, a Goth. When Theodosius died at the close of the fourth eentury he left two sons. Alaric supported one of these, Arcadius, in Constantinople, and Stilicho the other, Honorius, in Italy. In other words Alaric and Stilicho fought for the empire with the princes as puppets. In the course of their struggle Alaric marched into Italy and after a short siege took Rome (410 A.D.).

The opening half of the fifth century saw the whole of the Roman Empire in Europe the prey of robber armies of barbarians. It is difficult to visualize the state of affairs in the world at that time. Over France, Spain, Italy and the Balkan peninsula, the great cities that had flourished under the early empire still stood, impoverished, partly depopulated and falling into decay. Life in them must have been shallow, mean and full of uncertainty. Local officials asserted their authority and went on with their work with such conscience as they had, no doubt in the name of a now remote and inaccessible emperor. The churches went on, but usually with illiterate priests. There was little reading and much superstition and fear. But everywhere except where looters had destroyed them, books and pictures and statuary and such-like works of art were still to be found.

The life of the countryside had also degenerated. Everywhere this Roman world was much more weedy and untidy than it had been. In some regions war and pestilence had brought the land down to the level of a waste. Roads and forests were infested with robbers. Into such regions the barbarians marched, with little or no opposition, and set up their chiefs as rulers, often with Roman official titles. If they were half civilized barbarians they would give the conquered districts tolerable terms, they would take possession of the towns, associate and intermarry, and acquire (with an accent) the Latin speech; but the Jutes, the Angles and Saxons who submerged the Roman province of Britain were agriculturalists and had no use for towns, they seem to have swept south Britain clear of the Romanized population and they replaced the language by their own Teutonic dialects, which became at last English.

It is impossible in the space at our disposal to trace the movements of all the various German and Slavonic tribes as they went to and fro in the disorganized empire in search of plunder and a pleasant home. But let the Vandals serve as an example. They came into

Photo: Sebah & Joaillier

BASE OF THE "OBELISK OF THEODOSIUS," CONSTANTINOPLE

The obelisk of Thothmes, taken from Egypt to Constantinople by Theodosius and placed upon the pedestal here shown: an interesting example of early Byzantine art. The complete obelisk is seen on page 239.

history in east Germany. They settled as we have told in Pannonia. Thence they moved somewhen about 425 A.D. through the intervening provinces to Spain. There they found Visigoths from South Russia and other German tribes setting up dukes and kings. From Spain the Vandals under Genseric sailed for North Africa (429), captured Carthage (439), and built a fleet. They secured the mastery of the sea and captured and pillaged Rome (455), which had recovered very imperfectly from her capture and looting by Alaric half a century earlier. Then the Vandals made themselves masters of Sicily, Corsica, Sardinia and most of the other islands of the western Mediterranean. They made, in fact, a sea empire very similar in its extent to the sea empire of Carthage seven hundred odd years before. They were at the climax of their power about 477. They were a mere handful of conquerors holding all this country. In the next century almost all their territory had been reconquered for the empire of Constantinople during a transitory blaze of energy under Justinian I.

The story of the Vandals is but one sample of a host of similar adventures. But now there was coming into the European world the least kindred and most redoubtable of all these devastators, the Mongolian Huns or Tartars, a yellow people active and able, such as the western world had never before encountered.

XL

The Huns and the End of the Western Empire

THIS appearance of a conquering Mongolian people in Europe may be taken to mark a new stage in human history. Until the last century or so before the Christian era, the Mongol and the Nordic peoples had not been in close touch. Far away in the frozen lands beyond the northern forests the Lapps, a Mongolian people, had drifted westward as far as Lapland, but they played no part in the main current of history. For thousands of years the western world carried on the dramatic interplay of the Aryan, Semitic and fundamental brunette peoples with very little interference (except for an Ethiopian invasion of Egypt or so) either from the black peoples to the south or from the Mongolian world in the far East.

It is probable that there were two chief causes for the new westward drift of the nomadic Mongolians. One was the consolidation of the great empire of China, its extension northward and the increase of its population during the prosperous period of the Han dynasty. The other was some process of climatic change; a lesser rainfall that abolished swamps and forests perhaps, or a greater rainfall that extended grazing over desert steppes, or even perhaps both these processes going on in different regions but which anyhow facilitated a westward migration. A third contributary cause was the economic wretchedness, internal decay and falling population of the Roman Empire. The rich men of the later Roman Republic, and then the tax-gatherers of the military emperors had utterly consumed its vitality. So we have the factors of thrust, means and opportunity. There was pressure from the east, rot in the west and an open road.

The Hun had reached the eastern boundaries of European Russia by the first century A.D., but it was not until the fourth and fifth centuries A.D. that these horsemen rose to predominance upon the steppes. The fifth century was the Hun's century. The first Huns to come into Italy were mercenary bands in the pay of Stilicho the Vandal, the master of Honorius. Presently they were in possession of Pannonia, the empty nest of the Vandals.

By the second quarter of the fifth century a great war chief had arisen among the Huns, Attila. We have only vague and tantalizing glimpses of his power. He ruled not only over the Huns but over a conglomerate of tributary Germanic tribes; his empire extended from the Rhine cross the plains into Central Asia. He exchanged ambassadors with China. His head camp was in the plain of Hungary east of the Danube. There he was visited by an envoy from Constantinople, Priscus, who has left us an account of his state. The way of living of these Mongols was very like the way of living of the primitive Aryans they had replaced. The common folk were in huts and tents; the chiefs lived in great stockaded timber halls. There were feasts and drinking and singing by the bards. The Homeric heroes and even the Macedonian companions of Alexander would probably have felt more at home in the camp-capital of Attila than they would have done in the cultivated and decadent court of Theodosius II, the son of Arcadius, who was then reigning in Constantinople.

For a time it seemed as though the nomads under the leadership of the Huns and Attila would play the same part towards the Græco-Roman civilization of the Mediterranean countries that the barbaric Greeks had played long ago to the Ægean civilization. It looked like history repeating itself upon a larger stage. But the Huns were much more wedded to the nomadic life than the early Greeks, who were rather migratory cattle farmers than true nomads. The Huns raided and plundered but did not settle.

For some years Attila bullied Theodosius as he chose. His armies devastated and looted right down to the walls of Constantinople, Gibbon says that he totally destroyed no less than seventy cities in the Balkan peninsula, and Theodosius bought him off by payments of tribute and tried to get rid of him for good by sending secret agents to assassinate him. In 451 Attila turned his attention to the remains of the Latin-speaking half of the empire and invaded

Gaul. Nearly every town in northern Gaul was sacked. Franks, Visigoths and the imperial forces united against him and he was defeated at Troyes in a vast dispersed battle in which a multitude of men, variously estimated as between 150,000 and 300,000, were killed. This checked him in Gaul, but it did not exhaust his enormous military resources. Next year he came into Italy by way of Venetia, burnt Aquileia and Padua and looted Milan.

HEAD OF BARBARIAN CHIEF
(In the British Museum)

Numbers of fugitives from these north Italian towns and particularly from Padua fled to islands in the lagoons at the head of the Adriatic and laid there the foundations of the city state of Venice, which was to become one of the greatest of the trading centres in the middle ages.

In 453 Attila died suddenly after a great feast to celebrate his marriage to a young woman, and at his death this plunder confederation of his fell to pieces. The actual Huns disappear from history, mixed into the surrounding more numerous Aryan-speaking populations. But these great Hun raids practically consummated the end of the Latin Roman Empire. After his death ten different emperors ruled in Rome in twenty years, set up by Vandal and other mercenary troops. The Vandals from Carthage took and sacked Rome in 455. Finally in 476 Odoacer, the chief of the barbarian troops, suppressed a Pannonian who was figuring as emperor under the impressive name of Romulus Augustulus, and informed the Court of Constantinople that there was no longer an emperor in the west. So ingloriously the Latin Roman Empire came to an end. In 493 Theodoric the Goth became King of Rome.

All over western and central Europe now barbarian chiefs were reigning as kings, dukes and the like, practically independent but for the most part professing some sort of shadowy allegiance to the emperor. There were hundreds and perhaps thousands of such practically independent brigand rulers. In Gaul, Spain and Italy and in Dacia the Latin speech still prevailed in locally distorted forms, but in Britain and east of the Rhine languages of the German group (or in Bohemia a Slavonic language, Czech) were the common speech. The superior clergy and a small remnant of other educated men read and wrote Latin. Everywhere life was insecure and property was held by the strong arm. Castles multiplied and roads fell into decay. The dawn of the sixth century was an age of division and of intellectual darkness throughout the western world. Had it not been for the monks and Christian missionaries Latin learning might have perished altogether.

Why had the Roman Empire grown and why had it so completely decayed? It grew because at first the idea of citizenship held it together. Throughout the days of the expanding republic, and even into the days of the early empire there remained a great number of men conscious of Roman citizenship, feeling it a privilege and an obligation to be a Roman citizen, confident of their rights under the Roman law and willing to make sacrifices in the name of Rome. The prestige of Rome as of something just and great and law-upholding spread far beyond the Roman boundaries. But even as early as the Punic wars the sense of citizenship was being undermined by the growth of wealth and slavery. Citizenship spread indeed but not the idea of citizenship.

The Roman Empire was after all a very primitive organization; it did not educate, did not explain itself to its increasing multitudes of citizens, did not invite their co-operation in its decisions. There was no network of schools to ensure a common understanding, no distribution of news to sustain collective activity. The adventurers who struggled for power from the days of Marius and Sulla onward had no idea of creating and calling in public opinion upon the imperial affairs. The spirit of citizenship died of starvation and no one observed it die. All empires, all states, all organizations of human society are, in the ultimate, things of understanding and will. There remained no will for the Roman Empire in the World and so it came to an end.

But though the Latin-speaking Roman Empire died in the fifth century, something else had been born within it that was to avail itself enormously of its prestige and tradition, and that was the Latin-speaking half of the Catholic Church. This lived while the empire died because it appealed to the minds and wills of men, because it had books and a great system of teachers and missionaries to hold it together, things stronger than any law or legions. Throughout the fourth and fifth centuries A.D. while the empire was decaying, Christianity was spreading to a universal dominion in Europe. It conquered its conquerors, the barbarians. When Attila seemed disposed to march on Rome, the patriarch of Rome intercepted him and did what no armies could do, turning him back by sheer moral force.

The Patriarch or Pope of Rome claimed to be the head of the entire Christian church. Now that there were no more emperors, he began to annex imperial titles and claims. He took the title of *pontifex maximus*, head sacrificial priest of the Roman dominion, the most ancient of all the titles that the emperors had enjoyed.

XLI

The Byzantine and Sassanid Empires

THE Greek-speaking eastern half of the Roman Empire showed much more political tenacity than the western half. It weathered the disasters of the fifth century A.D., which saw a complete and final breaking up of the original Latin Roman power. Attila bullied the Emperor Theodosius II and sacked and raided almost to the walls of Constantinople, but that city remained intact. The Nubians came down the Nile and looted Upper Egypt, but Lower Egypt and Alexandria were left still fairly prosperous. Most of Asia Minor was held against the Sassanid Persians.

The sixth century, which was an age of complete darkness for the West, saw indeed a considerable revival of the Greek power. Justinian I (527–565) was a ruler of very great ambition and energy, and he was married to the Empress Theodora, a woman of quite equal capacity who had begun life as an actress. Justinian reconquered North Africa from the Vandals and most of Italy from the Goths. He even regained the south of Spain. He did not limit his energies to naval and military enterprises. He founded a university, built the great church of Sta. Sophia in Constantinople and codified the Roman law. But in order to destroy a rival to his university foundation he closed the schools of philosophy in Athens, which had been going on in unbroken continuity from the days of Plato, that is to say for nearly a thousand years.

From the third century onwards the Persian Empire had been the steadfast rival of the Byzantine. The two empires kept Asia Minor, Syria and Egypt in a state of perpetual unrest and waste. In the first century A.D., these lands were still at a high level of civilization, wealthy and with an abundant population, but the continual coming and going of armies, massacres, looting and war taxation wore them down steadily until only shattered and ruinous cities remained upon a countryside of scattered peasants. In this melancholy process of impoverishment and disorder lower Egypt fared perhaps less badly than the rest of the world. Alexandria, like Constantinople, continued a dwindling trade between the east and the west.

Science and political philosophy seemed dead now in both these warring and decaying empires. The last philosophers of Athens, until their suppression, preserved the texts of the great literature of the past with an infinite reverence and want of understanding. But there remained no class of men in the world, no free gentlemen with bold and independent habits of thought, to carry on the tradition of frank statement and enquiry embodied in these writings. The social and political chaos accounts largely for the disappearance of this class, but there was also another reason why the human intelligence was sterile and feverish during this age. In both Persia and Byzantium it was an age of intolerance. Both empires were religious empires in a new way, in a way that greatly hampered the free activities of the human mind.

Photo: Sebah & Joaillier

THE CHURCH (NOW A MOSQUE) OF S. SOPHIA, CONSTANTINOPLE
The obelisk of Theodosius is in the foreground

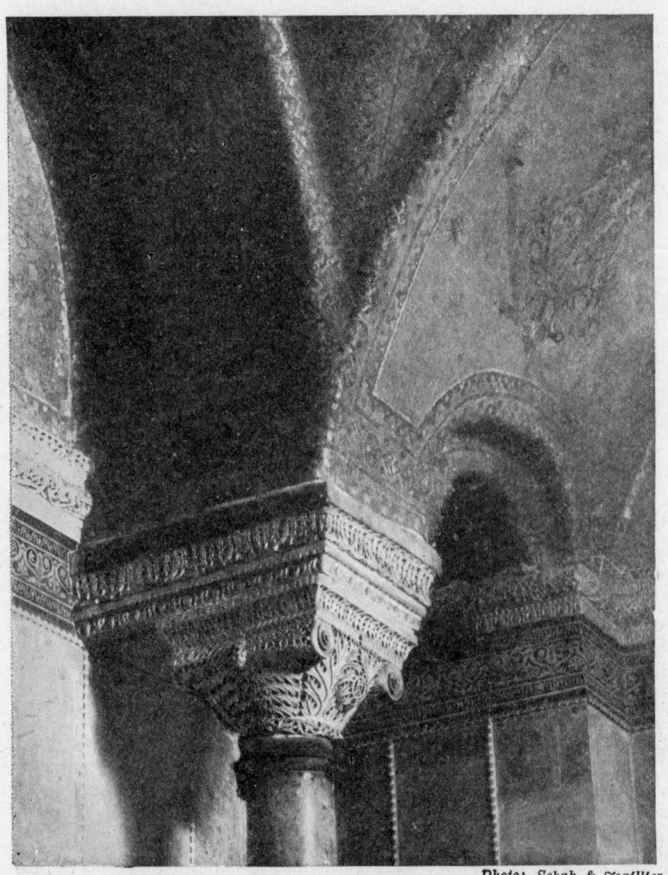

Photo: Sebah & Joaillier

THE MAGNIFICENT ROOF-WORK IN S. SOPHIA

Of course the oldest empires in the world were religious empires, centring upon the worship of a god or of a god-king. Alexander was treated as a divinity and the Cæsars were gods in so much as they had altars and temples devoted to them and the offering of incense was made a test of loyalty to the Roman state. But these older religions were essentially religions of act and fact. They did not invade the mind. If a man offered his sacrifice and bowed to the god, he was left not only to think but to say practically whatever he liked about the affair. But the new sort of religions that had come into the world, and particularly Christianity, turned inward. These new faiths demanded not simply conformity but understanding belief. Naturally fierce controversy ensued upon the exact meaning of the things believed. These new religions were creed religions. The world was confronted with a new word, Orthodoxy, and with a stern resolve to keep not only acts but speech and private thought within the limits of a set teaching. For to hold a wrong opinion, much more to convey it to other people, was no longer regarded as an intellectual defect but a moral fault that might condemn a soul to everlasting destruction.

Both Ardashir I who founded the Sassanid dynasty in the third century A.D., and Constantine the Great who reconstructed the Roman Empire in the fourth, turned to religious organizations for help, because in these organizations they saw a new means of using and controlling the wills of men. And already before the end of the fourth century both empires were persecuting free talk and religious innovation. In Persia Ardashir found the ancient Persian religion of Zoroaster (or Zarathushtra) with its priests and temples and a sacred fire that burnt upon its altars, ready for his purpose as a state religion. Before the end of the third century Zoroastrianism was persecuting Christianity, and in 277 A.D. Mani, the founder of

Photo: Alinari

THE RAVENNA PANEL, DEPICTING JUSTINIAN AND HIS COURT

Photo: Underwood & Underwood

THE ROCK–HEWN TEMPLE AT PETRA

a new faith, the Manichæans, was crucified and his body flayed. Constantinople, on its side, was busy hunting out Christian heresies. Manichæan ideas infected Christianity and had to be fought with the fiercest methods; in return ideas from Christianity affected the purity of the Zoroastrian doctrine. All ideas became suspect. Science, which demands before all things the free action of an untroubled mind, suffered a complete eclipse throughout this phase of intolerance.

War, the bitterest theology, and the usual vices of mankind constituted Byzantine life of those days. It was picturesque, it was romantic; it had little sweetness or light. When Byzantium and Persia were not fighting the barbarians from the north, they wasted Asia Minor and Syria in dreary and destructive hostilities. Even in close alliance these two empires would have found it a hard task to turn back the barbarians and recover their prosperity. The Turks or Tartars first come into history as the allies first of one power and then of another. In the sixth century the two chief antagonists were Justinian and Chosroes I; in the opening of the seventh the Emperor Heraclius was pitted against Chosroes II (580).

At first and until after Heraclius had become Emperor (610) Chosroes II carried all before him. He took Antioch, Damascus and Jerusalem and his armies reached Chalcedon, which is in Asia Minor over against Constantinople. In 619 he conquered Egypt. Then Heraclius pressed a counter attack home and routed a Persian army at Nineveh (627), although at that time there were still Persian troops at Chalcedon. In 628 Chosroes II was deposed and murdered by his son, Kavadh, and an inconclusive peace was made between the two exhausted empires.

Byzantium and Persia had fought their last war. But few people as yet dreamt of the storm that was even then gathering in the deserts to put an end for ever to this aimless, chronic struggle.

While Heraclius was restoring order in Syria a message reached him. It had been brought in to the imperial outpost at Bostra south of Damascus; it was in Arabic, an obscure Semitic desert language, and it was read to the Emperor, if it reached him at all, by an interpreter. It was from someone who called himself "Muhammad the Prophet of God." It called upon the Emperor to acknowledge the One True God and to serve him. What the Emperor said is not recorded.

A similar message came to Kavadh at Ctesiphon. He was annoyed, tore up the letter, and bade the messenger begone.

This Muhammad, it appeared, was a Bedouin leader whose headquarters were in the mean little desert town of Medina. He was preaching a new religion of faith in the One True God.

"Even so, O Lord!" he said; "rend thou his Kingdom from Kavadh."

XLII

The Dynasties of Suy and Tang in China

THROUGHOUT the fifth, sixth, seventh and eighth centuries, there was a steady drift of Mongolian peoples westward. The Huns of Attila were merely precursors of this advance, which led at last to the establishment of Mongolian peoples in Finland, Esthonia, Hungary and Bulgaria, where their descendants, speaking languages akin to Turkish, survive to this day. The Mongolian nomads were, in fact, playing a rôle towards the Aryanized civilizations of Europe and Persia and India that the Aryans had played to the Ægean and Semitic civilizations ten or fifteen centuries before.

In Central Asia the Turkish peoples had taken root in what is now Western Turkestan, and Persia already employed many Turkish officials and Turkish mercenaries. The Parthians had gone out of history, absorbed into the general population of Persia. There were no more Aryan nomads in the history of Central Asia; Mongolian people had replaced them. The Turks became masters of Asia from China to the Caspian.

The same great pestilence at the end of the second century A.D. that had shattered the Roman Empire had overthrown the Han dynasty in China. Then came a period of division and of Hunnish conquests from which China arose refreshed, more rapidly and more completely than Europe was destined to do. Before the end of the sixth century China was reunited under the Suy dynasty, and this by the time of Heraclius gave place to the Tang dynasty, whose reign marks another great period of prosperity for China.

Throughout the seventh, eighth and ninth centuries China was the most secure and civilized country in the world. The Han dynasty had extended her boundaries in the north; the Suy and Tang dynasties now spread her civilization to the south, and China

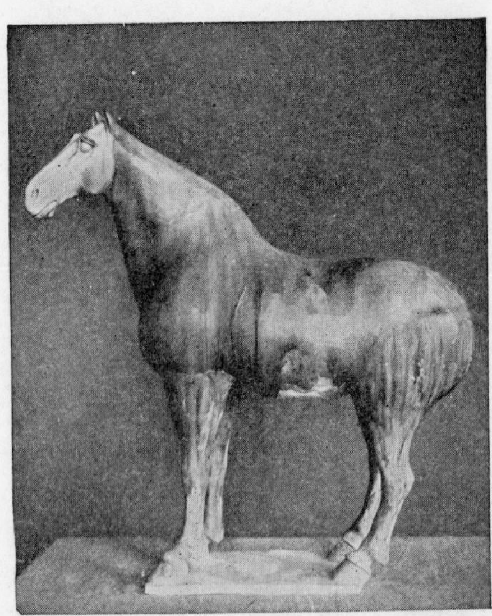
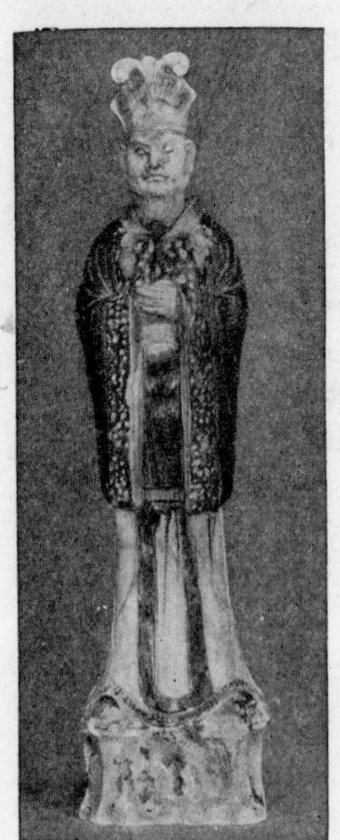

CHINESE EARTHENWARE ART OF THE TANG DYNASTY, 618–906

Specimens in glazed earthenware, in brown, green and buff, discovered in tombs in China

(In the Victoria and Albert Museum)

began to assume the proportions she has to-day. In Central Asia indeed she reached much further, extending at last, through tributary Turkish tribes, to Persia and the Caspian Sea.

The new China that had arisen was a very different land from the old China of the Hans. A new and more vigorous literary school appeared, there was a great poetic revival; Buddhism had revolutionized philosophical and religious thought. There were great advances in artistic work, in technical skill and in all the amenities of life. Tea was first used, paper manufactured and wood-block printing began. Millions of people indeed were leading orderly, graceful and kindly lives in China during these centuries when the attenuated populations of Europe and Western Asia were living either in hovels, small walled cities or grim robber fortresses. While the mind of the west was black with theological obsessions, the mind of China was open and tolerant and enquiring.

One of the earliest monarchs of the Tang dynasty was Tai-tsung, who began to reign in 627, the year of the victory of Heraclius at Nineveh. He received an embassy from Heraclius, who was probably seeking an ally in the rear of Persia. From Persia itself came a party of Christian missionaries (635). They were allowed to explain their creed to Tai-tsung and he examined a Chinese translation of their Scriptures. He pronounced this strange religion acceptable, and gave permission for the foundation of a church and monastery.

To this monarch also (in 628) came messengers from Muhammad. They came to Canton on a trading ship. They had sailed the whole way from Arabia along the Indian coasts. Unlike Heraclius and Kavadh, Tai-tsung gave these envoys a courteous hearing. He expressed his interest in their theological ideas and assisted them to build a mosque in Canton, a mosque which survives, it is said, to this day, the oldest mosque in the world.

XLIII

Muhammad and Islam

A PROPHETIC amateur of history surveying the world in the opening of the seventh century might have concluded very reasonably that it was only a question of a few centuries before the whole of Europe and Asia fell under Mongolian domination. There were no signs of order or union in Western Europe, and the Byzantine and Persian Empires were manifestly bent upon a mutual destruction. India also was divided and wasted. On the other hand China was a steadily expanding empire which probably at that time exceeded all Europe in population, and the Turkish people who were growing to power in Central Asia were disposed to work in accord with China. And such a prophecy would not have been an altogether vain one. A time was to come in the thirteenth century when a Mongolian overlord would rule from the Danube to the Pacific, and Turkish dynasties were destined to reign over the entire Byzantine and Persian Empires, over Egypt and most of India.

Where our prophet would have been most likely to have erred would have been in under-estimating the recuperative power of the Latin end of Europe and in ignoring the latent forces of the Arabian desert. Arabia would have seemed what it had been for times immemorial, the refuge of small and bickering nomadic tribes. No Semitic people had founded an empire now for more than a thousand years.

Then suddenly the Bedouin flared out for a brief century of splendour. They spread their rule and language from Spain to the boundaries of China. They gave the world a new culture. They created a religion that is still to this day one of the most vital forces in the world.

The man who fired this Arab flame appears first in history as the young husband of the widow of a rich merchant of the town of Mecca, named Muhammad. Until he was forty he did very little to distinguish himself in the world. He seems to have taken considerable interest in religious discussion. Mecca was a pagan city at that time worshipping in particular a black stone, the Kaaba, of great repute throughout all Arabia and a centre of pilgrimages; but there were great numbers of Jews in the country — indeed all the southern portion of Arabia professed the Jewish faith — and there were Christian churches in Syria.

About forty Muhammad began to develop prophetic characteristics like those of the Hebrew prophets twelve hundred years before him. He talked first to his wife of the One True God, and of the rewards and punishments of virtue and wickedness. There can be no doubt that his thoughts were very strongly influenced by Jewish and Christian ideas. He gathered about him a small circle of believers and presently began to preach in the town against the prevalent idolatry. This made him extremely unpopular with his fellow townsmen because the pilgrimages to the Kaaba were the chief source of such prosperity as Mecca enjoyed. He became bolder and more definite in his teaching, declaring himself to be the last chosen prophet of God entrusted with a mission to perfect religion. Abraham, he declared, and Jesus Christ were his forerunners. He had been chosen to complete and perfect the revelation of God's will.

He produced verses which he said had been communicated to him by an angel, and he had a strange vision in which he was taken up through the Heavens to God and instructed in his mission.

As his teaching increased in force the hostility of his fellow townsmen increased also. At last a plot was made to kill him; but he escaped with his faithful friend and disciple, Abu Bekr, to the friendly town of Medina which adopted his doctrine. Hostilities followed between Mecca and Medina which ended at last in a treaty. Mecca was to adopt the worship of the One True God and accept Muhummad as his prophet, *but the adherents of the new faith were still to make the pilgrimage to Mecca* just as they had done when they were pagans. So Muhammad established the One True God in

Photo: Lehnert & Landrock

AT PRAYER IN THE DESERT

Mecca without injuring its pilgrim traffic. In 629 Muhammad returned to Mecca as its master, a year after he had sent out these envoys of his to Heraclius, Tai-tsung, Kavadh and all the rulers of the earth.

Then for four years more until his death in 632, Muhammad spread his power over the rest of Arabia. He married a number of wives in his declining years, and his life on the whole was by modern standards unedifying. He seems to have been a man compounded of very considerable vanity, greed, cunning, self-deception and quite

LOOKING ACROSS THE SEA OF SAND *Photo: Lehnert & Landrock*

sincere religious passion. He dictated a book of injunctions and expositions, the Koran, which he declared was communicated to him from God. Regarded as literature or philosophy the Koran is certainly unworthy of its alleged Divine authorship.

Yet when the manifest defects of Muhammad's life and writings have been allowed for, there remains in Islam, this faith he imposed upon the Arabs, much power and inspiration. One is its uncompromising monotheism; its simple enthusiastic faith in the rule and fatherhood of God and its freedom from theological complications. Another is its complete detachment from the sacrificial priest and the temple. It is an entirely prophetic religion, proof against any possibility of relapse towards blood sacrifices. In the Koran the limited and ceremonial nature of the pilgrimage to Mecca is stated beyond the possibility of dispute, and every precaution was taken by Muhammad to prevent the deification of himself after his death. And a third element of strength lay in the insistence of Islam upon the perfect brotherhood and equality before God of all believers, whatever their colour, origin or status.

These are the things that made Islam a power in human affairs. It has been said that the true founder of the Empire of Islam was not so much Muhammad as his friend and helper, Abu Bekr. If Muhammad, with his shifty character, was the mind and imagination of primitive Islam, Abu Bekr was its conscience and its will. Whenever Muhammad wavered Abu Bekr sustained him. And when Muhammad died, Abu Bekr became Caliph (= successor), and with that faith that moves mountains, he set himself simply and sanely to organize the subjugation of the whole world to Allah — with little armies of 3,000 or 4,000 Arabs — according to those letters the prophet had written from Medina in 628 to all the monarchs of the world.

XLIV

The Great Days of the Arabs

THERE follows the most amazing story of conquest in the whole history of our race. The Byzantine army was smashed at the battle of the Yarmuk (a tributary of the Jordan) in 634; and the Emperor Heraclius, his energy sapped by dropsy and his resources exhausted by the Persian war, saw his new conquests in Syria, Damascus, Palmyra, Antioch, Jerusalem and the rest fall almost without resistance to the Moslim. Large elements in the population went over to Islam. Then the Moslim turned east. The Persians had found an able general in Rustam; they had a great host with a force of elephants; and for three days they fought the Arabs at Kadessia (637) and broke at last in headlong rout.

The conquest of all Persia followed, and the Moslem Empire pushed far into Western Turkestan and eastward until it met the Chinese. Egypt fell almost without resistance to the new conquerors, who full of a fanatical belief in the sufficiency of the Koran, wiped out the vestiges of the book-copying industry of the Alexandria Library. The tide of conquest poured along the north coast of Africa to the Straits of Gibraltar and Spain. Spain was invaded in 710 and the Pyrenees Mountains were reached in 720. In 732 the Arab advance had reached the centre of France, but here it was stopped for good at the battle of Poitiers and thrust back as far as the Pyrenees again. The conquest of Egypt had given the Moslim a fleet, and for a time it looked as though they would take Constantinople. They made repeated sea attacks between 672 and 718 but the great city held out against them.

The Arabs had little political aptitude and no political experience, and this great empire with its capital now at Damascus, which stretched from Spain to China, was destined to break up very speedily. From the very beginning doctrinal differences under-

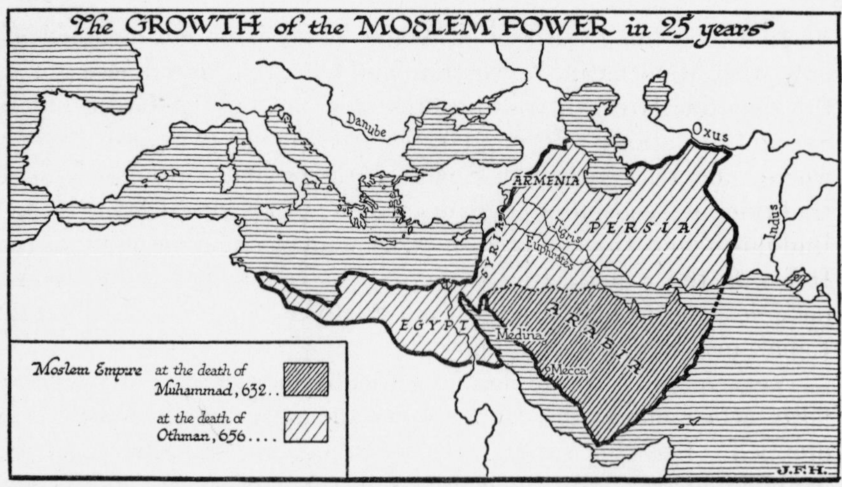

mined its unity. But our interest here lies not with the story of its political disintegration but with its effect upon the human mind and upon the general destinies of our race. The Arab intelligence had been flung across the world even more swiftly and dramatically than had the Greek a thousand years before. The intellectual stimulation of the whole world west of China, the break-up of old ideas and development of new ones, was enormous.

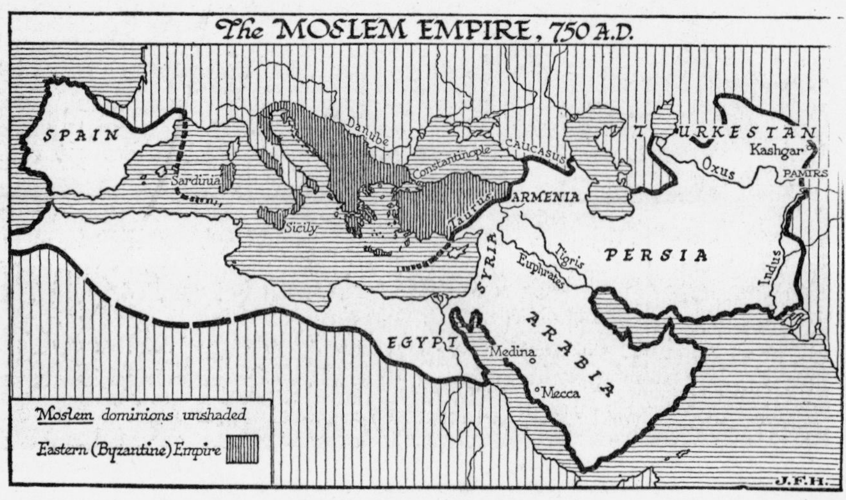

In Persia this fresh excited Arabic mind came into contact not only with Manichæan, Zoroastrian and Christian doctrine, but with the scientific Greek literature, preserved not only in Greek but in Syrian translations. It found Greek learning in Egypt also. Everywhere, and particularly in Spain, it discovered an active Jewish tradition of speculation and discussion. In Central Asia it met Buddhism and the material achievements of Chinese civilization. It learnt the manufacture of paper — which made printed books possible — from the Chinese. And finally it came into touch with Indian mathematics and philosophy.

Very speedily the intolerant self-sufficiency of the early days of faith, which made the Koran seem the only possible book, was dropped. Learning sprang up everywhere in the footsteps of the Arab conquerors. By the eighth century there was an educational

Photo: Lehnert & Landrock

JERUSALEM. SHOWING THE MOSQUE OF OMAR

VIEW OF CAIRO MOSQUES

Photo: Lehnert & Landrock

organization throughout the whole "Arabized" world. In the ninth learned men in the schools of Cordoba in Spain were corresponding with learned men in Cairo, Bagdad, Bokhara and Samarkand. The Jewish mind assimilated very readily with the Arab, and for a time the two Semitic races worked together through the medium of Arabic. Long after the political break-up and enfeeblement of the Arabs, this intellectual community of the Arab-speaking world endured. It was still producing very considerable results in the thirteenth century.

So it was that the systematic accumulation and criticism of facts which was first begun by the Greeks was resumed in this astonishing renascence of the Semitic world. The seed of Aristotle and the museum of Alexandria that had lain so long inactive and neglected now germinated and began to grow towards fruition. Very great advances were made in mathematical, medical and physical science.

The clumsy Roman numerals were ousted by the Arabic figures we use to this day and the zero sign was first employed. The very name *algebra* is Arabic. So is the word chemistry. The names of such stars as Algol, Aldebaran and Boötes preserve the traces of Arab conquests in the sky. Their philosophy was destined to reanimate the medieval philosophy of France and Italy and the whole Christian world.

The Arab experimental chemists were called alchemists, and they were still sufficiently barbaric in spirit to keep their methods and results secret as far as possible. They realized from the very beginning what enormous advantages their possible discoveries might give them, and what far-reaching consequences they might have on human life. They came upon many metallurgical and technical devices of the utmost value, alloys and dyes, distilling, tinctures and essences, optical glass; but the two chief ends they sought, they sought in vain. One was "the philosopher's stone"—a means of changing the metallic elements one into another and so getting a control of artificial gold, and the other was the *elixir vitæ*, a stimulant that would revivify age and prolong life indefinitely. The crabbed patient experimenting of these Arab alchemists spread into the Christian world. The fascination of their enquiries spread. Very gradually the activities of these alchemists became more social and co-operative. They found it profitable to exchange and compare ideas. By insensible gradations the last of the alchemists became the first of the experimental philosophers.

The old alchemists sought the philosopher's stone which was to transmute base metals to gold, and an elixir of immortality; they found the methods of modern experimental science which promise in the end to give man illimitable power over the world and over his own destiny.

XLV

The Development of Latin Christendom

IT is worth while to note the extremely shrunken dimensions of the share of the world remaining under Aryan control in the seventh and eighth centuries. A thousand years before, the Aryan-speaking races were triumphant over all the civilized world west of China. Now the Mongol had thrust as far as Hungary, nothing of Asia remained under Aryan rule except the Byzantine dominions in Asia Minor, and all Africa was lost and nearly all Spain. The great Hellenic world had shrunken to a few possessions round the nucleus of the trading city of Constantinople, and the memory of the Roman world was kept alive by the Latin of the western Christian priests. In vivid contrast to this tale of retrogression, the Semitic tradition had risen again from subjugation and obscurity after a thousand years of darkness.

Yet the vitality of the Nordic peoples was not exhausted. Confined now to Central and North-Western Europe and terribly muddled in their social and political ideas, they were nevertheless building up gradually and steadily a new social order and preparing unconsciously for the recovery of a power even more extensive than that they had previously enjoyed.

We have told how at the beginning of the sixth century there remained no central government in Western Europe at all. That world was divided up among numbers of local rulers holding their own as they could. This was too insecure a state of affairs to last; a system of co-operation and association grew up in this disorder, the feudal system, which has left its traces upon European life up to the present time. This feudal system was a sort of crystallization of society about power. Everywhere the lone man felt insecure and was prepared to barter a certain amount of his liberty for help and protection. He sought a stronger man as his lord and protector;

he gave him military services and paid him dues, and in return he was confirmed in his possession of what was his. His lord again found safety in vassalage to a still greater lord. Cities also found it convenient to have feudal protectors, and monasteries and church estates bound themselves by similar ties. No doubt in many cases allegiance was claimed before it was offered; the system grew downward as well as upward. So a sort of pyramidal system grew up, varying widely in different localities, permitting at first a considerable play of violence and private warfare but making steadily for order and a new reign of law. The pyramids grew up until some became recognizable as kingdoms. Already by the early sixth century a Frankish kingdom existed under its founder Clovis in what is now France and the Netherlands, and presently Visigothic and Lombard and Gothic kingdoms were in existence.

The Moslim when they crossed the Pyrenees in 720 found this Frankish kingdom under the practical rule of Charles Martel, the Mayor of the Palace of a degenerate descendant of Clovis, and experienced the decisive defeat of Poitiers (732) at his hands. This Charles Martel was practically overlord of Europe north of the Alps from the Pyrenees to Hungary. He ruled over a multitude of subordinate lords speaking French-Latin, and High and Low German languages. His son Pepin extinguished the last descendants of Clovis and took the kingly state and title. His grandson Charlemagne, who began to reign in 768, found himself lord of a realm so large that he could think of reviving the title of Latin Emperor. He conquered North Italy and made himself master of Rome.

Approaching the story of Europe as we do from the wider horizons of a world history we can see much more distinctly than the mere nationalist historian how cramping and disastrous this tradition of the Latin Roman Empire was. A narrow intense struggle for this phantom predominance was to consume European energy for more than a thousand years. Through all that period it is possible to trace certain unquenchable antagonisms; they run through the wits of Europe like the obsessions of a demented mind. One driving force was this ambition of successful rulers, which Charlemagne (Charles the Great) embodied, to become Cæsar. The realm of Charlemagne consisted of a complex of feudal German states at various stages of barbarism. West of the Rhine, most of these German peoples had learnt to speak various Latinized dialects which fused at last to form French. East of the Rhine, the racially similar German peoples did not lose their German speech. On account of this, communication was difficult between these two groups of barbarian conquerors and a split easily brought about. The split was made the more easy by the fact that the Frankish usage made it seem natural to divide the empire of Charlemagne among his sons at his death. So one aspect of the history of Europe from the days of Charlemagne onwards is a history of first this monarch and his family and then that, struggling to a precarious headship of the kings, princes, dukes, bishops and cities of Europe, while a steadily deepening antagonism between the French and German speaking elements develops in the medley. There was a formality of election for each emperor; and the climax of his ambition was to struggle to the possession of that worn-out, misplaced capital Rome and to a coronation there.

The next factor in the European political disorder was the resolve of the Church at Rome to make no temporal prince but the Pope of Rome himself emperor in effect. He was already pontifex maximus; for all practical purposes he held the decaying city; if he had no armies he had at least a vast propaganda organization in his priests throughout the whole Latin world; if he had little power over men's bodies he held the keys of heaven and hell in their imaginations and could exercise much influence upon their souls. So throughout the middle ages while one prince manœuvred against another first for equality, then for ascendancy, and at last for the supreme prize, the Pope of Rome, sometimes boldly, sometimes craftily, sometimes feebly — for the Popes were a succession of oldish men and the average reign of a Pope was not more than two years — manœuvred for the submission of all the princes to himself as the ultimate overlord of Christendom.

But these antagonisms of prince against prince and of Emperor against Pope do not by any means exhaust the factors of the European confusion. There was still an Emperor in Constantinople speaking Greek and claiming the allegiance of all Europe. When Charlemagne sought to revive the empire, it was merely the Latin end of the empire he revived. It was natural that a sense of rivalry between Latin Empire and Greek Empire should develop very readily. And still more readily did the rivalry of Greek-speaking Christianity and the newer Latin-speaking version develop. The Pope of Rome claimed to be the successor of St. Peter, the chief of the apostles of Christ, and the head of the Christian community everywhere. Neither the emperor nor the patriarch in Constantinople were disposed to acknowledge this claim. A dispute about a fine point in the doctrine of the Holy Trinity consummated a long series of dissensions in a final rupture in 1054. The Latin Church and the Greek Church became and remained thereafter distinct and frankly antagonistic. This antagonism must be added to the others in our estimate of the conflicts that wasted Latin Christendom in the middle ages.

Upon this divided world of Christendom rained the blows of three sets of antagonists. About the Baltic and North Seas remained a series of Nordic tribes who were only very slowly and reluc-

Photo: Rischgitz

STATUE OF CHARLEMAGNE IN FRONT OF NOTRE DAME, PARIS

The figure is entirely imaginary and romantic. There is no contemporary portrait of Charlemagne

tantly Christianized; these were the Northmen. They had taken to the sea and piracy, and were raiding all the Christian coasts down to Spain. They had pushed up the Russian rivers to the desolate central lands and brought their shipping over into the south-flowing rivers. They had come out upon the Caspian and Black Seas as pirates also. They set up principalities in Russia; they were the first people to be called Russians. These Northmen Russians came near to taking Constantinople. England in the early ninth century was a Christianized Low German country under a king, Egbert, a protégé and pupil of Charlemagne. The Northmen wrested half the kingdom from his successor Alfred the Great (886), and finally under Canute (1016) made themselves masters of the whole land. Under Rolph the Ganger (912) another band of Northmen conquered the north of France, which became Normandy.

Canute ruled not only over England but over Norway and Denmark, but his brief empire fell to pieces at his death through that political weakness of the barbaric peoples — division among a ruler's sons. It is interesting to speculate what might have happened if this temporary union of the Northmen had endured. They were a race of astonishing boldness and energy. They sailed in their galleys even to Iceland and Greenland. They were the first Europeans to land on American soil. Later on Norman adventurers were to recover Sicily from the Saracens and sack Rome. It is a fascinating thing to imagine what a great northern sea-faring power might have grown out of Canute's kingdom, reaching from America to Russia.

To the east of the Germans and Latinized Europeans was a medley of Slav tribes and Turkish peoples. Prominent among these were the Magyars or Hungarians who were coming westward throughout the eighth and ninth centuries. Charlemagne held them for a time, but after his death they established themselves in what is now Hungary; and after the fashion of their kindred predecessors, the Huns, raided every summer into the settled parts of Europe. In 938 they went through Germany into France, crossed the Alps into North Italy, and so came home, burning, robbing and destroying.

Finally pounding away from the south at the vestiges of the

Roman Empire were the Saracens. They had made themselves largely masters of the sea; their only formidable adversaries upon the water were the Northmen, the Russian Northmen out of the Black Sea and the Northmen of the west.

Hemmed in by these more vigorous and aggressive peoples, amidst forces they did not understand and dangers they could not

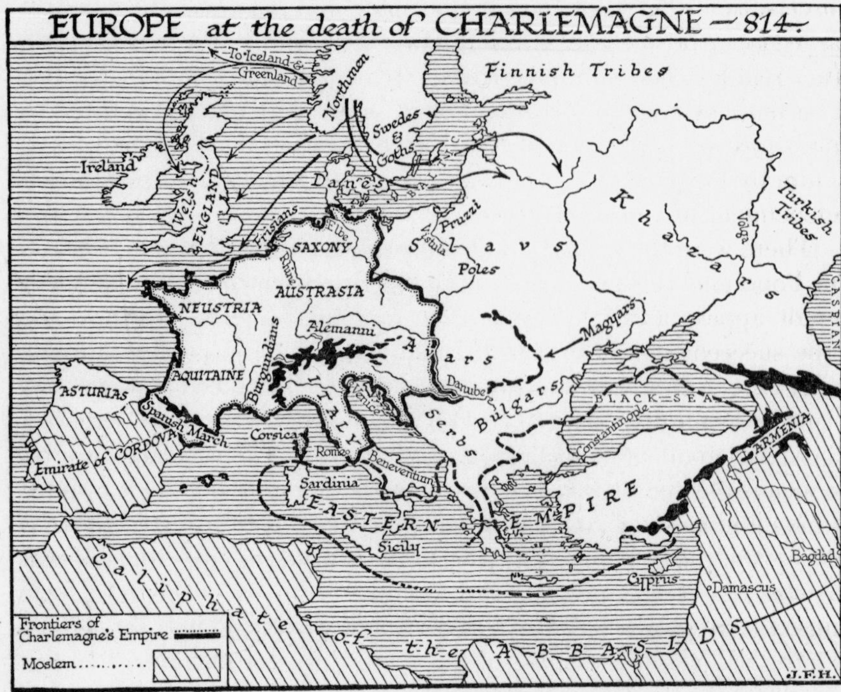

EUROPE at the death of CHARLEMAGNE — 814.

estimate, Charlemagne and after him a series of other ambitious spirits took up the futile drama of restoring the Western Empire under the name of the Holy Roman Empire. From the time of Charlemagne onward this idea obsessed the political life of Western Europe, while in the East the Greek half of the Roman power decayed and dwindled until at last nothing remained of it at all but the corrupt trading city of Constantinople and a few miles of territory about it. Politically the continent of Europe remained traditional and uncreative from the time of Charlemagne onward for a thousand years.

The name of Charlemagne looms large in European history but his personality is but indistinctly seen. He could not read nor write, but he had a considerable respect for learning; he liked to be read aloud to at meals and he had a weakness for theological discussion. At his winter quarters at Aix-la-Chapelle or Mayence he gathered about him a number of learned men and picked up much from their conversation. In the summer he made war, against the Spanish Saracens, against the Slavs and Magyars, against the Saxons, and other still heathen German tribes. It is doubtful whether the idea of becoming Cæsar in succession to Romulus Augustulus occurred to him before his acquisition of North Italy, or whether it was suggested to him by Pope Leo III, who was anxious to make the Latin Church independent of Constantinople.

There were the most extraordinary manœuvres at Rome between the Pope and the prospective emperor in order to make it appear or not appear as if the Pope gave him the imperial crown. The Pope succeeded in crowning his visitor and conqueror by surprise in St. Peter's on Christmas Day 800 A.D. He produced a crown, put it on the head of Charlemagne and hailed him Cæsar and Augustus. There was great applause among the people. Charlemagne was by no means pleased at the way in which the thing was done, it rankled in his mind as a defeat; and he left the most careful instructions to his son that he was not to let the Pope crown him emperor; he was to seize the crown into his own hands and put it on his own head himself. So at the very outset of this imperial revival we see beginning the age-long dispute of Pope and Emperor for priority. But Louis the Pious, the son of Charlemagne, disregarded his father's instructions and was entirely submissive to the Pope.

The empire of Charlemagne fell apart at the death of Louis the Pious and the split between the French-speaking Franks and the German-speaking Franks widened. The next emperor to arise was Otto, the son of a certain Henry the Fowler, a Saxon, who had been elected King of Germany by an assembly of German princes and prelates in 919. Otto descended upon Rome and was crowned emperor there in 962. This Saxon line came to an end early in the eleventh century and gave place to other German rulers. The feudal princes and nobles to the west who spoke various French dialects did not fall under the sway of these German emperors after the Carlovingian line, the line that is descended from Charlemagne, had come to an end, and no part of Britain ever came into the Holy Roman Empire. The Duke of Normandy, the King of France and a number of lesser feudal rulers remained outside.

In 987 the Kingdom of France passed out of the possession of the Carlovingian line into the hands of Hugh Capet, whose descendants were still reigning in the eighteenth century. At the time of Hugh Capet the King of France ruled only a comparatively small territory round Paris.

In 1066 England was attacked almost simultaneously by an invasion of the Norwegian Northmen under King Harold Hardrada and by the Latinized Northmen under the Duke of Normandy. Harold King of England defeated the former at the battle of Stamford Bridge, and was defeated by the latter at Hastings. England was conquered by the Normans, and so cut off from Scandinavian, Teutonic and Russian affairs, and brought into the most intimate relations and conflicts with the French. For the next four centuries the English were entangled in the conflicts of the French feudal princes and wasted upon the fields of France.

XLVI

The Crusades and the Age of Papal Dominion

IT is interesting to note that Charlemagne corresponded with the Caliph Haroun-al-Raschid, the Haroun-al-Raschid of the *Arabian Nights*. It is recorded that Haroun-al-Raschid sent ambassadors from Bagdad — which had now replaced Damascus as the Moslem capital — with a splendid tent, a water clock, an elephant and the keys of the Holy Sepulchre. This latter present was admirably calculated to set the Byzantine Empire and this new Holy Roman Empire by the ears as to which was the proper protector of the Christians in Jerusalem.

These presents remind us that while Europe in the ninth century was still a weltering disorder of war and pillage, there flourished a great Arab Empire in Egypt and Mesopotamia, far more civilized than anything Europe could show. Here literature and science still lived; the arts flourished, and the mind of man could move without fear or superstition. And even in Spain and North Africa where the Saracenic dominions were falling into political confusion there was a vigorous intellectual life. Aristotle was read and discussed by these Jews and Arabs during these centuries of European darkness. They guarded the neglected seeds of science and philosophy.

North-east of the Caliph's dominions was a number of Turkish tribes. They had been converted to Islam, and they held the faith much more simply and fiercely than the actively intellectual Arabs and Persians to the south. In the tenth century the Turks were growing strong and vigorous while the Arab power was divided and decaying. The relations of the Turks to the Empire of the Caliphate became very similar to the relations of the Medes to the last Babylonian Empire fourteen centuries before. In the eleventh century a group of Turkish tribes, the Seljuk Turks, came down into Mesopotamia and made the Caliph their nominal ruler but really their captive and tool. They conquered Armenia. Then they struck at the remnants of the Byzantine power in Asia Minor. In 1071 the Byzantine army was utterly smashed at the battle of Melasgird, and the Turks swept forward until not a trace of Byzantine rule remained in Asia. They took the fortress of Nicæa over against Constantinople, and prepared to attempt that city.

The Byzantine emperor, Michael VII, was overcome with terror. He was already heavily engaged in warfare with a band of Norman adventurers who had seized Durazzo, and with a fierce Turkish people, the Petschenegs, who were raiding over the Danube. In his extremity he sought help where he could, and it is notable that he did not appeal to the western emperor but to the Pope of Rome as the head of Latin Christendom. He wrote to Pope Gregory VII, and his successor Alexius Comnenus wrote still more urgently to Urban II.

This was not a quarter of a century from the rupture of the Latin and Greek churches. That controversy was still vividly alive in men's minds, and this disaster to Byzantium must have presented itself to the Pope as a supreme opportunity for reasserting the supremacy of the Latin Church over the dissentient Greeks. Moreover this occasion gave the Pope a chance to deal with two other matters that troubled western Christendom very greatly. One was the custom of "private war" which disordered social life, and the other was the superabundant fighting energy of the Low Germans and Christianized Northmen and particularly of the Franks and Normans. A religious war, the Crusade, the War of the Cross, was

Photo: Mansell

CRUSADER TOMBS IN EXETER CATHEDRAL

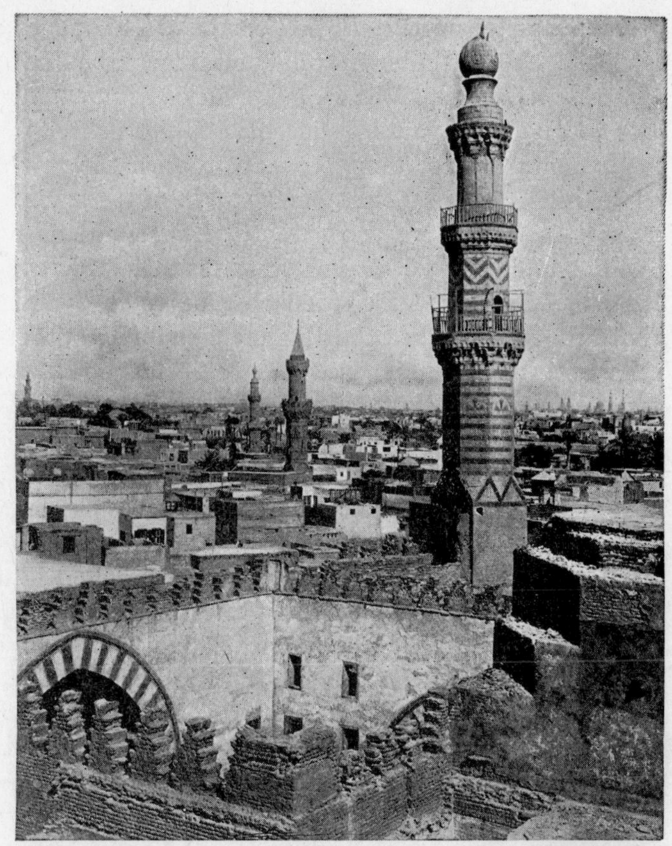

VIEW OF CAIRO

Photo: Lehnert & Landrock

preached against the Turkish captors of Jerusalem, and a truce to all warfare amongst Christians (1095). The declared object of this war was the recovery of the Holy Sepulchre from the unbelievers. A man called Peter the Hermit carried on a popular propaganda throughout France and Germany on broadly democratic lines. He went clad in a coarse garment, barefooted on an ass, he carried a huge cross and harangued the crowd in street or market-place or church. He denounced the cruelties practised upon the Christian pilgrims by the Turks, and the shame of the Holy Sepulchre being in any but Christian hands. The fruits of centuries of Christian teaching became apparent in the response. A great wave of enthusiasm swept the western world, and popular Christendom discovered itself.

Such a widespread uprising of the common people in relation to a single idea as now occurred was a new thing in the history of our race. There is nothing to parallel it in the previous history of the

Roman Empire or of India or China. On a smaller scale, however, there had been similar movements among the Jewish people after their liberation from the Babylonian captivity, and later on Islam was to display a parallel susceptibility to collective feeling. Such movements were certainly connected with the new spirit that had come into life with the development of the missionary-teaching religions. The Hebrew prophets, Jesus and his disciples, Mani, Muhammad, were all exhorters of men's individual souls. They brought the personal conscience face to face with God. Before that time religion had been much more a business of fetish, of pseudo-science, than of conscience. The old kind of religion turned upon temple, initiated priest and mystical sacrifice, and ruled the common man like a slave by fear. The new kind of religion made a man of him.

The preaching of the First Crusade was the first stirring of the common people in European history. It may be too much to call it the birth of modern democracy, but certainly at that time modern democracy stirred. Before very long we shall find it stirring again, and raising the most disturbing social and religious questions.

Certainly this first stirring of democracy ended very pitifully and lamentably. Considerable bodies of common people, crowds rather than armies, set out eastward from France and the Rhineland and Central Europe without waiting for leaders or proper equipment to rescue the Holy Sepulchre. This was the "people's crusade." Two great mobs blundered into Hungary, mistook the recently converted Magyars for pagans, committed atrocities and were massacred. A third multitude with a similarly confused mind, after a great pogrom of the Jews in the Rhineland, marched eastward, and was also destroyed in Hungary. Two other huge crowds, under the leadership of Peter the Hermit himself, reached Constantinople, crossed the Bosphorus, and were massacred rather than defeated by the Seljuk Turks. So began and ended this first movement of the European people, as people.

Next year (1097) the real fighting forces crossed the Bosphorus. Essentially they were Norman in leadership and spirit. They stormed Nicæa, marched by much the same route as Alexander had followed fourteen centuries before, to Antioch. The siege of Antioch kept them a year, and in June 1099 they invested Jerusalem. It was stormed after a month's siege. The slaughter was terrible. Men riding on horseback were splashed by the blood in the streets. At nightfall on July 15th the Crusaders had fought their way into the Church of the Holy Sepulchre and overcome all opposition there: blood-stained, weary and "sobbing from excess of joy" they knelt down in prayer.

Photo: D. McLeish

THE HORSES OF S. MARK, VENICE

Originally on the arch of Trajan at Constantinople, the Doge Dandalo V took them after the Fourth Crusade, to Venice, whence Napoleon I removed them to Paris, but in 1815 they were returned to Venice. During the Great War of 1914–18 they were hidden away for fear of air raids.

Immediately the hostility of Latin and Greek broke out again. The Crusaders were the servants of the Latin Church, and the Greek patriarch of Jerusalem found himself in a far worse case under the triumphant Latins than under the Turks. The Crusaders discovered themselves between Byzantine and Turk and fighting both. Much of Asia Minor was recovered by the Byzantine Empire, and the Latin princes were left, a buffer between Turk and Greek, with Jerusalem and a few small principalities, of which Edessa was one of the chief, in Syria. Their grip even on these possessions was precarious, and in 1144 Edessa fell to the Moslim, leading to an ineffective Second Crusade, which failed to recover Edessa but saved Antioch from a similar fate.

In 1169 the forces of Islam were rallied under a Kurdish adventurer named Saladin who had made himself master of Egypt. He preached a Holy War against the Christians, recaptured Jerusalem in 1187, and so provoked the Third Crusade. This failed to recover Jerusalem. In the Fourth Crusade (1202–4) the Latin Church turned frankly upon the Greek Empire, and there was not even a pretence of fighting the Turks. It started from Venice and in 1204 it stormed Constantinople. The great rising trading city of Venice was the leader in this adventure, and most of the coasts and islands of the Byzantine Empire were annexed by the Venetians. A "Latin" emperor (Baldwin of Flanders) was set up in Constantinople and the Latin and Greek Church were declared to be reunited. The Latin emperors ruled in Constantinople from 1204 to 1261 when the Greek world shook itself free again from Roman predominance.

The twelfth century then and the opening of the thirteenth was the age of papal ascendancy just as the eleventh was the age of the ascendancy of the Seljuk Turks and the tenth the age of the Northmen. A united Christendom under the rule of the Pope came nearer to being a working reality than it ever was before or after that time.

In those centuries a simple Christian faith was real and widespread over great areas of Europe. Rome itself had passed through some dark and discreditable phases; few writers can be found to excuse the lives of Popes John XI and John XII in the tenth century; they were abominable creatures; but the heart and body of Latin Christendom had remained earnest and simple; the generality of the common priests and monks and nuns had lived exemplary and faithful lives. Upon the wealth of confidence such lives created rested the power of the church. Among the great Popes of the past had been Gregory the Great, Gregory I (590–604) and Leo III (795–816) who invited Charlemagne to be Cæsar and crowned him in spite of himself. Towards the close of the eleventh century there arose a great clerical statesman, Hildebrand, who ended his life as Pope Gregory VII (1073–1085). Next but one after him came Urban II (1087–1099), the Pope of the First Crusade. These two were the founders of this period of papal greatness during which the Popes lorded it over the Emperors. From Bulgaria to Ireland and

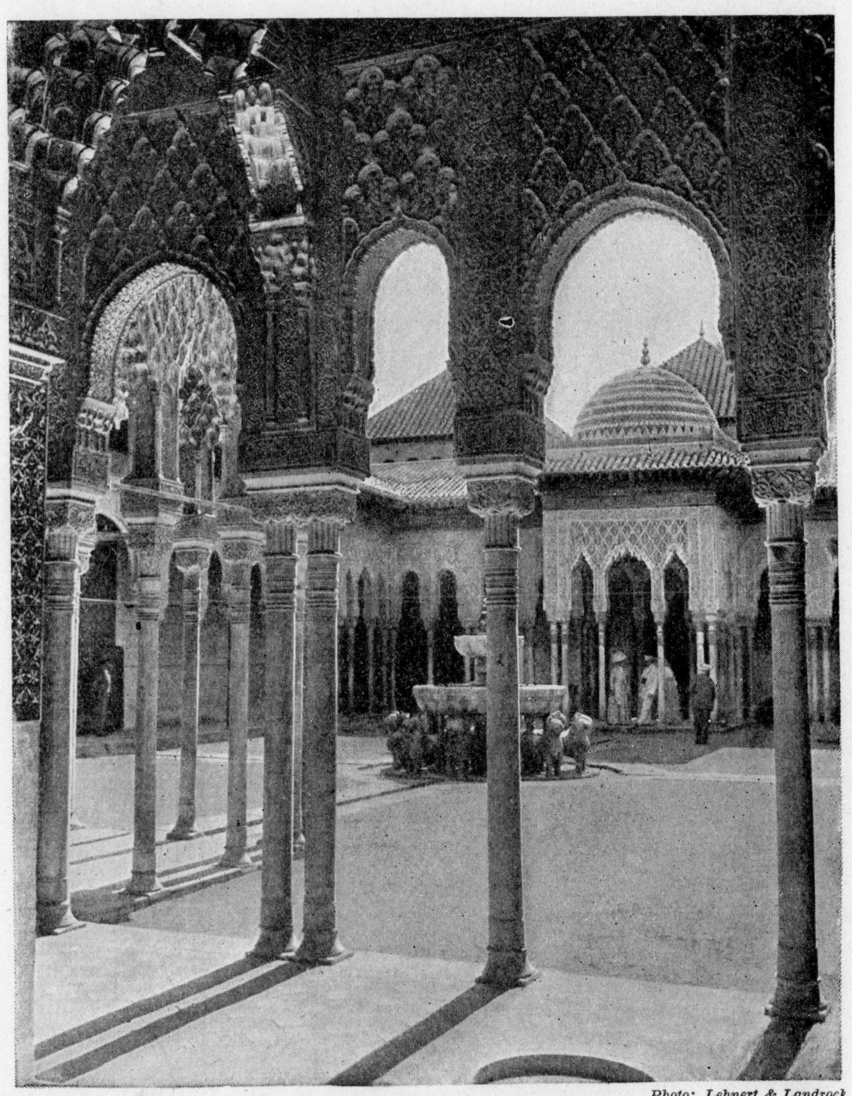

Photo: Lehnert & Landrock

A COURTYARD IN THE ALHAMBRA

from Norway to Sicily and Jerusalem the Pope was supreme. Gregory VII obliged the Emperor Henry IV to come in penitence to him at Canossa and to await forgiveness for three days and nights in the courtyard of the castle, clad in sackcloth and barefooted to the snow. In 1176 at Venice the Emperor Frederick (Frederick Barbarossa), knelt to Pope Alexander III and swore fealty to him.

The great power of the church in the beginning of the eleventh century lay in the wills and consciences of men. It failed to retain the moral prestige on which its power was based. In the opening decades of the fourteenth century it was discovered that the power of the Pope had evaporated. What was it that destroyed the naïve confidence of the common people of Christendom in the church so that they would no longer rally to its appeal and serve its purposes?

The first trouble was certainly the accumulation of wealth by the church. The church never died, and there was a frequent disposition on the part of dying childless people to leave lands to the church. Penitent sinners were exhorted to do so. Accordingly in many European countries as much as a fourth of the land became church property. The appetite for property grows with what it feeds upon. Already in the thirteenth century it was being said everywhere that the priests were not good men, that they were always hunting for money and legacies.

The kings and princes disliked this alienation of property very greatly. In the place of feudal lords capable of military support, they found their land supporting abbeys and monks and nuns. And these lands were really under foreign dominion. Even before the time of Pope Gregory VII there had been a struggle between the princes and the papacy over the question of "investitures," the question that is of who should appoint the bishops. If that power rested with the Pope and not the King, then the latter lost control not only of the consciences of his subjects but of a considerable part of his dominions. For also the clergy claimed exemption from taxation. They paid their taxes to Rome. And not only that, but the church also claimed the right to levy a tax of one-tenth upon the property of the layman in addition to the taxes he paid his prince.

The history of nearly every country in Latin Christendom tells of the same phase in the eleventh century, a phase of struggle between monarch and Pope on the issue of investitures and generally it tells of a victory for the Pope. He claimed to be able to excommunicate the prince, to absolve his subjects from their allegiance to him, to recognize a successor. He claimed to be able to put a nation under an interdict, and then nearly all priestly functions ceased except the sacraments of baptism, confirmation and penance; the priests could neither hold the ordinary services, marry people, nor bury the dead. With these two weapons it was possible for the twelfth century Popes to curb the most recalcitrant princes and overawe the most restive peoples. These were enormous powers, and enormous powers are only to be used on extraordinary occasions. The Popes used them at last with a frequency that staled their effect. Within thirty years at the end of the twelfth century we find Scotland, France and England in turn under an interdict. And also the Popes could not resist the temptation to preach crusades against offending princes — until the crusading spirit was extinct.

It is possible that if the Church of Rome had struggled simply against the princes and had had a care to keep its hold upon the general mind, it might have achieved a permanent dominion over all Christendom. But the high claims of the Pope were reflected as arrogance in the conduct of the clergy. Before the eleventh century the Roman priests could marry; they had close ties with the people among whom they lived; they were indeed a part of the people. Gregory VII made them celibates; he cut the priests off from too great an intimacy with the laymen in order to bind them more closely to Rome, but indeed he opened a fissure between the church and the commonalty. The church had its own law courts. Cases involving not merely priests but monks, students, crusaders, widows, orphans and the helpless were reserved for the clerical courts, and so were all matters relating to wills, marriages and oaths and all cases of sorcery, heresy and blasphemy. Whenever the layman found himself in conflict with the priest he had to go to a clerical court. The obligations of peace and war fell upon his shoulders alone and left the priest free. It is no great wonder that jealousy and hatred of the priests grew up in the Christian world.

Never did Rome seem to realize that its power was in the consciences of common men. It fought against religious enthusiasm, which should have been its ally, and it forced doctrinal orthodoxy upon honest doubt and aberrant opinion. When the church interfered in matters of morality it had the common man with it, but not when it interfered in matters of doctrine. When in the south of France Waldo taught a return to the simplicity of Jesus in faith and life, Innocent III preached a crusade against the Waldenses, Waldo's followers, and permitted them to be suppressed with fire, sword, rape and the most abominable cruelties. When again St. Francis of Assisi (1181–1226) taught the imitation of Christ and a life of poverty and service, his followers, the Franciscans, were persecuted, scourged, imprisoned and dispersed. In 1318 four of them were burnt alive at Marseilles. On the other hand the fiercely orthodox order of the Dominicans, founded by St. Dominic (1170–1221) was strongly supported by Innocent III, who with its assistance set up an organization, the Inquisition, for the hunting of heresy and the affliction of free thought.

So it was that the church by excessive claims, by unrighteous privileges, and by an irrational intolerance destroyed that free faith of the common man which was the final source of all its power. The story of its decline tells of no adequate foemen from without but continually of decay from within.

XLVII

Recalcitrant Princes and the Great Schism

ONE very great weakness of the Roman Church in its struggle to secure the headship of all Christendom was the manner in which the Pope was chosen.

If indeed the papacy was to achieve its manifest ambition and establish one rule and one peace throughout Christendom, then it was vitally necessary that it should have a strong, steady and continuous direction. In those great days of its opportunity it needed before all things that the Popes when they took office should be able men in the prime of life, that each should have his successor-designate with whom he could discuss the policy of the church, and that the forms and processes of election should be clear, definite, unalterable and unassailable. Unhappily none of these things obtained. It was not even clear who could vote in the election of a Pope, nor whether the Byzantine or Holy Roman Emperor had a voice in the matter. That very great papal statesman Hildebrand (Pope Gregory VII, 1073–1085) did much to regularize the election. He confined the votes to the Roman cardinals and he reduced the Emperor's share to a formula of assent conceded to him by the church, but he made no provision for a successor-designate and he left it possible for the disputes of the cardinals to keep the See vacant, as in some cases it was kept vacant, for a year or more.

The consequences of this want of firm definition are to be seen in the whole history of the papacy up to the sixteenth century. From quite early times onward there were disputed elections and two or more men each claiming to be Pope. The church would then be subjected to the indignity of going to the Emperor or some other outside arbiter to settle the dispute. And the career of every one of the great Popes ended in a note of interrogation. At his death the church might be left headless and as ineffective as a decapitated

MILAN CATHEDRAL

View showing the exquisite carvings characteristic of the 98 spires of the edifice

body. Or he might be replaced by some old rival eager only to discredit and undo his work. Or some enfeebled old man tottering on the brink of the grave might succeed him.

It was inevitable that this peculiar weakness of the papal organization should attract the interference of the various German princes, the French King, and the Norman and French Kings who ruled in England; that they should all try to influence the elections, and have a Pope in their own interest established in the Lateran Palace at Rome. And the more powerful and important the Pope became in European affairs, the more urgent did these interventions become. Under the circumstances it is no great wonder that many of the Popes were weak and futile. The astonishing thing is that many of them were able and courageous men.

One of the most vigorous and interesting of the Popes of this great period was Innocent III (1198–1216) who was so fortunate as to become Pope before he was thirty-eight. He and his successors were pitted against an even more interesting personality, the Emperor Frederick II; *Stupor mundi* he was called, the Wonder of the world. The struggle of this monarch against Rome is a turning place in history. In the end Rome defeated him and destroyed his dynasty, but he left the prestige of the church and Pope so badly wounded that its wounds festered and led to its decay.

Frederick was the son of the Emperor Henry VI and his mother was the daughter of Roger I, the Norman King of Sicily. He inherited this kingdom in 1198 when he was a child of four years. Innocent III had been made his guardian. Sicily in those days had been but recently conquered by the Normans; the Court was half oriental and full of highly educated Arabs; and some of these were associated in the education of the young king. No doubt they were at some pains to make their point of view clear to him. He got a Moslem view of Christianity as well as a Christian view of Islam, and the unhappy result of this double system of instruction was a view, exceptional in that age of faith, that all religions were impostures. He talked freely on the subject; his heresies and blasphemies are on record.

As the young man grew up he found himself in conflict with his guardian. Innocent III wanted altogether too much from his ward.

A TYPICAL CRUSADER: DON RODRIGO DE CARDENAS
From the Church of S. Pedro at Ocana, Spain
(*In the Victoria and Albert Museum*)

When the opportunity came for Frederick to succeed as Emperor, the Pope intervened with conditions. Frederick must promise to put down heresy in Germany with a strong hand. Moreover he must relinquish his crown in Sicily and South Italy, because otherwise he would be too strong for the Pope. And the German clergy were to be freed from all taxation. Frederick agreed—but with no intention of keeping his word. The Pope had already induced the French King to make war upon his own subjects in France, the cruel and bloody crusade against the Waldenses; he wanted Frederick to do the same thing in Germany. But Frederick being far more of a heretic than any of the simple pietists who had incurred the Pope's animosity, lacked the crusading impulse. And when Innocent urged him to crusade against the Moslim and recover Jerusalem he was equally ready to promise and equally slack in his performance.

Having secured the imperial crown Frederick II stayed in Sicily, which he greatly preferred to Germany as a residence, and did nothing to redeem any of his promises to Innocent III, who died baffled in 1216.

Honorius III, who succeeded Innocent, could do no better with Frederick, and Gregory IX (1227) came to the papal throne evidently resolved to settle accounts with this young man at any cost. He excommunicated him. Frederick II was denied all the comforts of religion. In the half-Arab Court of Sicily this produced singularly little discomfort. And also the Pope addressed a public letter to the Emperor reciting his vices (which were indisputable), his heresies, and his general misconduct. To this Frederick replied in a document of diabolical ability. It was addressed to all the princes of Europe, and it made the first clear statement of the issue between the Pope and the princes. He made a shattering attack upon the manifest ambition of the Pope to become the absolute ruler of all Europe. He suggested a union of princes against this usurpation. He directed the attention of the princes specifically to the wealth of the church.

Having fired off this deadly missile Frederick resolved to perform his twelve-year-old promise and go upon a crusade. This was the Sixth Crusade (1228). It was, as a crusade, farcical. Frederick II went to Egypt and met and discussed affairs with the Sultan. These two gentlemen, both of sceptical opinions, exchanged congenial views, made a commercial convention to their mutual advantage, and agreed to transfer Jerusalem to Frederick. This indeed was a new sort of crusade, a crusade by private treaty. Here was no blood splashing the conqueror, no "weeping with excess of joy." As this astonishing crusader was an excommunicated man, he had to be content with a purely secular coronation as King of Jerusalem, taking the crown from the altar with his own hand — for all the clergy were bound to shun him. He then returned to Italy, chased the papal armies which had invaded his dominions back to their own territories, and obliged the Pope to grant him absolution from his excommunication. So a prince might treat the Pope in the thirteenth century, and there was now no storm of popular indignation to avenge him. Those days were past.

In 1239 Gregory IX resumed his struggle with Frederick, excommunicated him for a second time, and renewed that warfare of public abuse in which the papacy had already suffered severely. The controversy was revived after Gregory IX was dead, when Innocent IV was Pope; and again a devastating letter, which men were bound to remember, was written by Frederick against the church. He denounced the pride and irreligion of the clergy, and ascribed all the corruptions of the time to their pride and wealth. He proposed to his fellow princes a general confiscation of church property — for the good of the church. It was a suggestion that never afterwards left the imagination of the European princes.

We will not go on to tell of his last years. The particular events of his life are far less significant than its general atmosphere. It is possible to piece together something of his court life in Sicily. He was luxurious in his way of living, and fond of beautiful things. He is described as licentious. But it is clear that he was a man of very effectual curiosity and inquiry. He gathered Jewish and Moslem as well as Christian philosophers at his court, and he did much to irrigate the Italian mind with Saracenic influences. Through him the Arabic numerals and algebra were introduced to Christian students, and among other philosophers at his court was Michael Scott, who translated portions of Aristotle and the commentaries thereon of the great Arab philosopher Averroes (of Cordoba). In 1224 Frederick founded the University of Naples, and he enlarged and enriched the great medical school at Salerno University. He also founded a zoological garden. He left a book on hawking, which shows him to have been an acute observer of the habits of birds, and he was one of the first Italians to write Italian verse. Italian poetry was indeed born at his court. He has been called by an able writer, "the first of the moderns," and the phrase expresses aptly the unprejudiced detachment of his intellectual side.

A still more striking intimation of the decay of the living and sustaining forces of the papacy appeared when presently the Popes came into conflict with the growing power of the French King. During the lifetime of the Emperor Frederick II, Germany fell into disunion, and the French King began to play the rôle of guard, supporter and rival to the Pope that had hitherto fallen to the Hohenstaufen Emperors. A series of Popes pursued the policy of supporting the French monarchs. French princes were established in the kingdom of Sicily and Naples, with the support and approval of Rome, and the French Kings saw before them the possibility of

COSTUMES OF THE BURGUNDIAN NOBILITY: FLEMISH WORK OF THE FIFTEENTH CENTURY

restoring and ruling the Empire of Charlemagne. When, however, the German interregnum after the death of Frederick II, the last of the Hohenstaufens, came to an end and Rudolf of Habsburg was elected first Habsburg Emperor (1273), the policy of Rome began to fluctuate between France and Germany, veering about with the sympathies of each successive Pope. In the East in 1261 the Greeks recaptured Constantinople from the Latin emperors, and the founder of the new Greek dynasty, Michael Palæologus, Michael VIII, after some unreal tentatives of reconciliation with the Pope, broke away from the Roman communion altogether, and with that, and the fall of the Latin kingdoms in Asia, the eastward ascendancy of the Popes came to an end

In 1294 Boniface VIII became Pope. He was an Italian, hostile to the French, and full of a sense of the great traditions and mission of Rome. For a time he carried things with a high hand. In 1300 he held a jubilee, and a vast multitude of pilgrims assembled in Rome. "So great was the influx of money into the papal treasury, that two assistants were kept busy with the rakes collecting the offerings that were deposited at the tomb of St. Peter."[1] But this festival was a delusive triumph. Boniface came into conflict with the French King in 1302, and in 1303, as he was about to pronounce sentence of excommunication against that monarch, he was surprised and arrested in his own ancestral palace at Anagni, by Guillaume de Nogaret. This agent from the French King forced an entrance into the palace, made his way into the bedroom of the frightened Pope — he was lying in bed with a cross in his hands — and heaped threats and insults upon him. The Pope was liberated a day or so later by the townspeople, and returned to Rome; but there he was seized upon and again made prisoner by the Orsini family, and in a few weeks' time the shocked and disillusioned old man died a prisoner in their hands.

The people of Anagni did resent the first outrage, and rose against Nogaret to liberate Boniface, but then Anagni was the Pope's native town. The important point to note is that the French King

[1] J. H. Robinson.

COSTUMES OF THE BURGUNDIAN NOBILITY: FLEMISH WORK OF THE FIFTEENTH CENTURY

This series is from casts in the Victoria and Albert Museum of the original brass statuettes in the Rijks Museum, Amsterdam

in this rough treatment of the head of Christendom was acting with the full approval of his people; he had summoned a council of the Three Estates of France (lords, church and commons) and gained their consent before proceeding to extremities. Neither in Italy, Germany nor England was there the slightest general manifestation of disapproval at this free handling of the sovereign pontiff. The idea of Christendom had decayed until its power over the minds of men had gone.

Throughout the fourteenth century the papacy did nothing to recover its moral sway. The next Pope elected, Clement V, was a Frenchman, the choice of King Philip of France. He never came to Rome. He set up his court in the town of Avignon, which then belonged not to France but to the papal See, though embedded in French territory, and there his successors remained until 1377, when Pope Gregory XI returned to the Vatican palace in Rome. But Gregory XI did not take the sympathies of the whole church with him. Many of the cardinals were of French origin and their habits and associations were rooted deep at Avignon. When in 1378 Gregory XI died, and an Italian, Urban VI, was elected, these dissentient cardinals declared the election invalid, and elected another Pope, the anti-Pope, Clement VII. This split is called the Great Schism. The Popes remained in Rome, and all the anti-French powers, the Emperor, the King of England, Hungary, Poland and the North of Europe were loyal to them. The anti-Popes, on the other hand, continued in Avignon, and were supported by the King of France, his ally the King of Scotland, Spain, Portugal and various German princes. Each Pope excommunicated and cursed the adherents of his rival (1378–1417).

Is it any wonder that presently all over Europe people began to think for themselves in matters of religion?

The beginnings of the Franciscans and the Dominicans, which we have noted in the preceding chapters, were but two among many of the new forces that were arising in Christendom, either to hold or shatter the church as its own wisdom might decide. Those two orders the church did assimilate and use, though with a little violence in the case of the former. But other forces were more frankly disobedient and critical. A century and a half later came Wycliffe (1320–1384). He was a learned Doctor at Oxford. Quite late in his life he began a series of outspoken criticisms of the corruption of the clergy and the unwisdom of the church. He organized a number of poor priests, the Wycliffites, to spread his ideas throughout England; and in order that people should judge between the church and himself, he translated the Bible into English. He was a more learned and far abler man than either St. Francis or St. Dominic. He had supporters in high places and a great following among the people; and though Rome raged against him, and ordered his imprisonment, he died a free man. But the black and ancient spirit that was leading the Catholic Church to its destruction would not let his bones rest in the grave. By a decree of the Council of Constance in 1415, his remains were ordered to be dug up and burnt, an order which was carried out at the command of Pope Martin V by Bishop Fleming in 1428. This desecration was not the act of some isolated fanatic; it was the official act of the church.

XLVIII

The Mongol Conquests

But in the thirteenth century, while this strange and finally ineffectual struggle to unify Christendom under the rule of the Pope was going on in Europe, far more momentous events were afoot upon the larger stage of Asia. A Turkish people from the country to the north of China rose suddenly to prominence in the world's affairs, and achieved such a series of conquests as has no parallel in history. These were the Mongols. At the opening of the thirteenth century they were a horde of nomadic horsemen, living very much as their predecessors, the Huns, had done, subsisting chiefly upon meat and mare's milk and living in tents of skin. They had shaken themselves free from Chinese dominion, and brought a number of other Turkish tribes into a military confederacy. Their central camp was at Karakorum in Mongolia.

At this time China was in a state of division. The great dynasty of Tang had passed into decay by the tenth century, and after a phase of division into warring states, three main empires, that of Kin in the north with Pekin as its capital and that of Sung in the south with a capital at Nankin, and Hsia in the centre, remain. In 1214 Jengis Khan, the leader of the Mongol confederates, made war on the Kin Empire and captured Pekin (1214). He then turned westward and conquered Western Turkestan, Persia, Armenia, India down to Lahore, and South Russia as far as Kieff. He died master of a vast empire that reached from the Pacific to the Dnieper.

His successor, Ogdai Khan, continued this astonishing career of conquest. His armies were organized to a very high level of efficiency; and they had with them a new Chinese invention, gunpowder, which they used in small field guns. He completed the conquest of the Kin Empire and then swept his hosts right across Asia to Russia (1235), an altogether amazing march. Kieff was

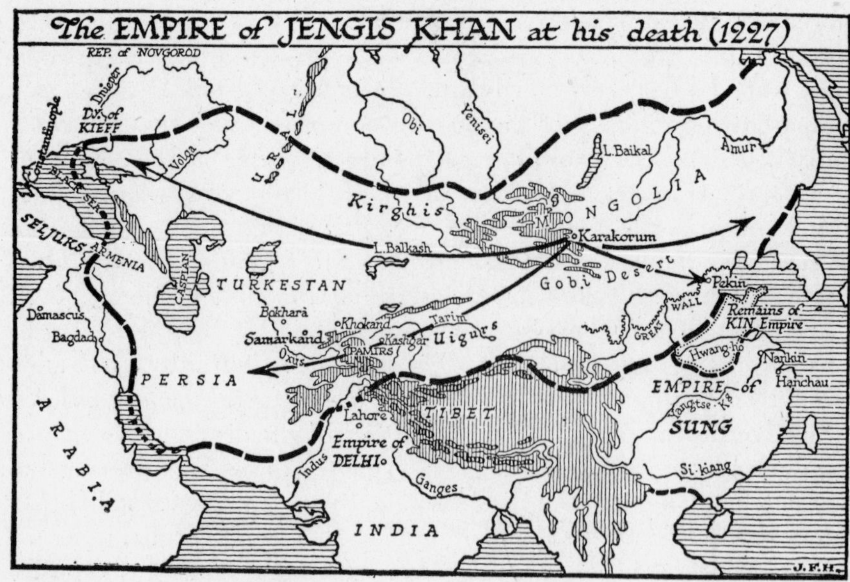

destroyed in 1240, and nearly all Russia became tributary to the Mongols. Poland was ravaged, and a mixed army of Poles and Germans was annihilated at the battle of Liegnitz in Lower Silesia in 1241. The Emperor Frederick II does not seem to have made any great efforts to stay the advancing tide.

"It is only recently," says Bury in his notes to Gibbon's *Decline and Fall of the Roman Empire*, "that European history has begun to understand that the successes of the Mongol army which overran Poland and occupied Hungary in the spring of A.D. 1241 were won by consummate strategy and were not due to a mere overwhelming superiority of numbers. But this fact has not yet become a matter of common knowledge; the vulgar opinion which represents the Tartars as a wild horde carrying all before them solely by their multitude, and galloping through Eastern Europe without a strategic plan, rushing at all obstacles and overcoming them by mere weight, still prevails. . . .

"It was wonderful how punctually and effectually the arrangements were carried out in operations extending from the Lower Vistula to Transylvania. Such a campaign was quite beyond the power of any European army of the time, and it was beyond the vision of any European commander. There was no general in Europe, from Frederick II downward, who was not a tyro in strategy compared to Subutai. It should also be noticed that the Mongols embarked upon the enterprise with full knowledge of the political situation of Hungary and the condition of Poland — they had taken care to inform themselves by a well-organized system of spies; on the other hand, the Hungarians and the Christian powers, like childish barbarians, knew hardly anything about their enemies."

But though the Mongols were victorious at Liegnitz, they did not continue their drive westward. They were getting into woodlands and hilly country, which did not suit their tactics; and so they turned southward and prepared to settle in Hungary, massacring or assimilating the kindred Magyar, even as these had previously massacred and assimilated the mixed Scythians and Avars and Huns before them. From the Hungarian plain they would probably have made raids west and south as the Hungarians had done in the ninth century, the Avars in the seventh and eighth and the Huns in the fifth. But Ogdai died suddenly, and in 1242 there was trouble

about the succession, and recalled by this, the undefeated hosts of Mongols began to pour back across Hungary and Roumania towards the east.

Thereafter the Mongols concentrated their attention upon their Asiatic conquests. By the middle of the thirteenth century they had conquered the Sung Empire. Mangu Khan succeeded Ogdai Khan as Great Khan in 1251, and made his brother Kublai Khan governor of China. In 1280 Kublai Khan had been formally recognized Emperor of China, and so founded the Yuan dynasty which lasted until 1368. While the last ruins of the Sung rule were going down in China, another brother of Mangu, Hulagu, was conquering Persia and Syria. The Mongols displayed a bitter animosity to Islam at this time, and not only massacred the population of Bagdad when they captured that city, but set to work to destroy the immemorial irrigation system which had kept Mesopotamia incessantly prosperous and populous from the early days of Sumeria. From that time until our own Mesopotamia has been a desert of ruins, sustaining only a scanty population. Into Egypt the Mongols never penetrated; the Sultan of Egypt completely defeated an army of Hulagu's in Palestine in 1260.

After that disaster the tide of Mongol victory ebbed. The dominions of the Great Khan fell into a number of separate states. The eastern Mongols became Buddhists, like the Chinese; the western became Moslim. The Chinese threw off the rule of the Yuan dynasty in 1368, and set up the native Ming dynasty which flourished from 1368 to 1644. The Russians remained tributary to the Tartar hordes upon the south-east steppes until 1480, when the Grand Duke of Moscow repudiated his allegiance and laid the foundation of modern Russia.

In the fourteenth century there was a brief revival of Mongol vigour under Timurlane, a descendant of Jengis Khan. He established himself in Western Turkestan, assumed the title of Grand Khan in 1369, and conquered from Syria to Delhi. He was the most savage and destructive of all the Mongol conquerors. He established an empire of desolation that did not survive his death. In 1505, however, a descendant of this Timur, an adventurer named Baber, got together an army with guns and swept down upon the

TARTAR HORSEMEN

(From a Chinese Print in the British Museum)

plains of India. His grandson Akbar (1556–1605) completed his conquests, and this Mongol (or "Mogul" as the Arabs called it) dynasty ruled in Delhi over the greater part of India until the eighteenth century.

One of the consequences of the first great sweep of Mongol conquest in the thirteenth century was to drive a certain tribe of Turks, the Ottoman Turks, out of Turkestan into Asia Minor. They extended and consolidated their power in Asia Minor, crossed the

Dardanelles and conquered Macedonia, Serbia and Bulgaria, until at last Constantinople remained like an island amongst the Ottoman dominions. In 1453 the Ottoman Sultan, Muhammad II, took Constantinople, attacking it from the European side with a great number of guns. This event caused intense excitement in Europe and there was talk of a crusade, but the day of the crusades was past.

In the course of the sixteenth century the Ottoman Sultans conquered Bagdad, Hungary, Egypt and most of North Africa, and their fleet made them masters of the Mediterranean. They very nearly took Vienna, and they exacted a tribute from the Emperor. There were but two items to offset the general ebb of Christian dominion in the fifteenth century. One was the restoration of the independence of Moscow (1480); the other was the gradual reconquest of Spain by the Christians. In 1492, Granada, the last Moslem state in the peninsula, fell to King Ferdinand of Aragon and his Queen Isabella of Castile.

But it was not until as late as 1571 that the naval battle of Lepanto broke the pride of the Ottomans, and restored the Mediterranean waters to Christian ascendancy.

XLIX

The Intellectual Revival of the Europeans

THROUGHOUT the twelfth century there were many signs that the European intelligence was recovering courage and leisure, and preparing to take up again the intellectual enterprises of the first Greek scientific enquiries and such speculations as those of the Italian Lucretius. The causes of this revival were many and complex. The suppression of private war, the higher standards of comfort and security that followed the crusades, and the stimulation of men's minds by the experiences of these expeditions were no doubt necessary preliminary conditions. Trade was reviving; cities were recovering ease and safety; the standard of education was arising in the church and spreading among laymen. The thirteenth and fourteenth centuries were a period of growing, independent or quasi-independent cities; Venice, Florence, Genoa, Lisbon, Paris, Bruges, London, Antwerp, Hamburg, Nuremberg, Novgorod, Wisby and Bergen for example. They were all trading cities with many travellers, and where men trade and travel they talk and think. The polemics of the Popes and princes, the conspicuous savagery and wickedness of the persecution of heretics, were exciting men to doubt the authority of the church and question and discuss fundamental things.

We have seen how the Arabs were the means of restoring Aristotle to Europe, and how such a prince as Frederick II acted as a channel through which Arabic philosophy and science played upon the renascent European mind. Still more influential in the stirring up of men's ideas were the Jews. Their very existence was a note of interrogation to the claims of the church. And finally the secret, fascinating enquiries of the alchemists were spreading far and wide and setting men to the petty, furtive and yet fruitful resumption of experimental science.

And the stir in men's minds was by no means confined now to the independent and well educated. The mind of the common man was awake in the world as it had never been before in all the experience of mankind. In spite of priest and persecution, Christianity does seem to have carried a mental ferment wherever its teaching reached. It established a direct relation between the conscience of the individual man and the God of Righteousness, so that now if need arose he had the courage to form his own judgment upon prince or prelate or creed.

As early as the eleventh century philosophical discussion had begun again in Europe, and there were great and growing universities at Paris, Oxford, Bologna and other centres. There medieval "schoolmen" took up again and thrashed out a series of questions upon the value and meaning of words that were a necessary preliminary to clear thinking in the scientific age that was to follow. And standing by himself because of his distinctive genius was Roger Bacon (circa 1210 to circa 1293), a Franciscan of Oxford, the father of modern experimental science. His name deserves a prominence in our history second only to that of Aristotle.

His writings are one long tirade against ignorance. He told his age it was ignorant, an incredibly bold thing to do. Nowadays a man may tell the world it is as silly as it is solemn, that all its methods are still infantile and clumsy and its dogmas childish assumptions, without much physical danger; but these peoples of the middle ages when they were not actually being massacred or starving or dying of pestilence, were passionately convinced of the wisdom, the completeness and finality of their beliefs, and disposed to resent any reflections upon them very bitterly. Roger Bacon's writings were like a flash of light in a profound darkness. He combined his attack upon the ignorance of his times with a wealth of suggestion for the increase of knowledge. In his passionate insistence upon the need of experiment and of collecting knowledge, the spirit of Aristotle lives again in him. "Experiment, experiment," that is the burthen of Roger Bacon.

Yet of Aristotle himself Roger Bacon fell foul. He fell foul of him because men, instead of facing facts boldly, sat in rooms and pored over the bad Latin translations which were then all that was available of the master. "If I had my way," he wrote, in his intemperate fashion, "I should burn all the books of Aristotle, for the study of them can only lead to a loss of time, produce error, and increase ignorance," a sentiment that Aristotle would probably have echoed could he have returned to a world in which his works were not so much read as worshipped — and that, as Roger Bacon showed, in these most abominable translations.

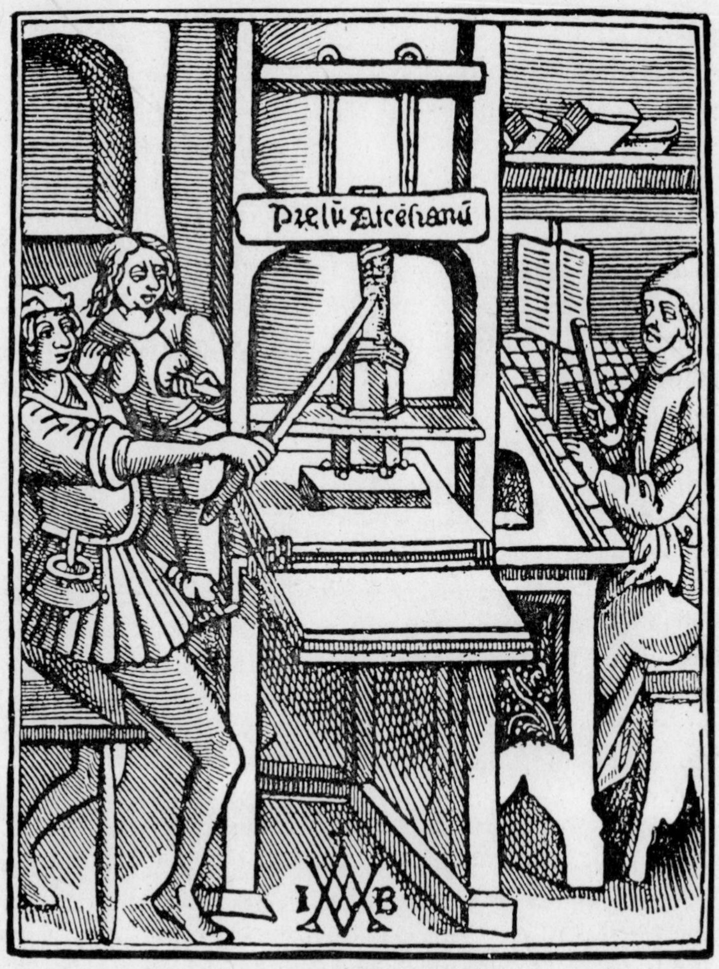

AN EARLY PRINTING PRESS

(From an old print)

Throughout his books, a little disguised by the necessity of seeming to square it all with orthodoxy for fear of the prison and worse, Roger Bacon shouted to mankind, "Cease to be ruled by dogmas and authorities; *look at the world!*" Four chief sources of ignorance he denounced; respect for authority, custom, the sense of the ignorant crowd, and the vain, proud unteachableness of our dispositions. Overcome but these, and a world of power would open to men:—

"Machines for navigating are possible without rowers, so that great ships suited to river or ocean, guided by one man, may be borne with greater speed than if they were full of men. Likewise cars may be made so that without a draught animal they may be moved *cum impetu inœstimable*, as we deem the scythed chariots to have been from which antiquity fought. And flying machines are possible, so that a man may sit in the middle turning some device by which artificial wings may beat the air in the manner of a flying bird."

So Roger Bacon wrote, but three more centuries were to elapse before men began any systematic attempts to explore the hidden stores of power and interest he realized so clearly existed beneath the dull surface of human affairs.

But the Saracenic world not only gave Christendom the stimulus of its philosophers and alchemists; it also gave it paper. It is scarcely too much to say that paper made the intellectual revival of Europe possible. Paper originated in China, where its use probably goes back to the second century B.C. In 751 the Chinese made an attack upon the Arab Moslems in Samarkand; they were repulsed, and among the prisoners taken from them were some skilled papermakers, from whom the art was learnt. Arabic paper manuscripts from the ninth century onward still exist. The manufacture entered Christendom either through Greece or by the capture of Moorish paper-mills during the Christian reconquest of Spain. But under the Christian Spanish the product deteriorated sadly. Good paper was not made in Christian Europe until the end of the thirteenth century, and then it was Italy which led the world. Only by the fourteenth century did the manufacture reach Germany, and not until the end of that century was it abundant and cheap enough for the printing of books to be a practicable business proposition. Thereupon printing followed naturally and necessarily, for printing is the most obvious of inventions, and the intellectual life of the world entered upon a new and far more vigorous phase. It ceased to be a little trickle from mind to mind; it became a broad flood, in which thousands and presently scores and hundreds of thousands of minds participated.

One immediate result of this achievement of printing was the appearance of an abundance of Bibles in the world. Another was a cheapening of school-books. The knowledge of reading spread swiftly. There was not only a great increase of books in the world, but the books that were now made were plainer to read and so easier to understand. Instead of toiling at a crabbed text and then thinking over its significance, readers now could think unimpeded as they read. With this increase in the facility of reading, the reading public grew. The book ceased to be a highly decorated toy or a scholar's mystery. People began to write books to be read as well as looked at by ordinary people. They wrote in the ordinary language and not in Latin. With the fourteenth century the real history of the European literature begins.

So far we have been dealing only with the Saracenic share in the European revival. Let us turn now to the influence of the Mongol conquests. They stimulated the geographical imagination of Europe enormously. For a time under the Great Khan, all Asia and Western Europe enjoyed an open intercourse; all the roads were temporarily open, and representatives of every nation appeared at the court of Karakorum. The barriers between Europe and Asia set up by the religious feud of Christianity and Islam were lowered. Great hopes were entertained by the papacy for the conversion of the Mongols to Christianity. Their only religion so far had been Shumanism, a primitive paganism. Envoys of the Pope, Buddhist priests from India, Parisian and Italian and Chinese artificers, Byzantine and Armenian merchants, mingled with Arab officials and Persian and Indian astronomers and mathematicians at the Mongol court. We hear too much in history of the campaigns and massacres of the Mongols, and not enough of their curiosity and desire for learning. Not perhaps as an originative people, but as transmitters of knowledge and method their influence upon the world's history has been very great. And everything one can learn of the vague and romantic personalities of Jengis or Kublai tends to confirm the impression that these men were at least as understanding and creative monarchs as either that flamboyant but egotistical figure Alexander the Great or that raiser of political ghosts, that energetic but illiterate theologian Charlemagne.

ANCIENT BRONZE FIGURE FROM BENIN, W. AFRICA

Note evidence in attire of knowledge of early European explorers

(*In the British Museum*)

One of the most interesting of these visitors to the Mongol Court was a certain Venetian, Marco Polo, who afterwards set down his story in a book. He went to China about 1272 with his father and uncle, who had already once made the journey. The Great Khan had been deeply impressed by the elder Polos; they were the first men of the "Latin" peoples he had seen; and he sent them back with enquiries for teachers and learned men who could explain Christianity to him, and for various other European things that had aroused his curiosity. Their visit with Marco was their second visit.

The three Polos started by way of Palestine and not by the Crimea, as in their previous expedition. They had with them a gold tablet and other indications from the Great Khan that must have greatly facilitated their journey. The Great Khan had asked for some oil from the lamp that burns in the Holy Sepulchre at Jerusalem; and so thither they first went, and then by way of Cilicia into Armenia. They went thus far north because the Sultan of Egypt was raiding the Mongol domains at this time. Thence they came by way of Mesopotamia to Ormuz on the Persian Gulf, as if they contemplated a sea voyage. At Ormuz they met merchants from India. For some reason they did not take ship, but instead turned northward through the Persian deserts, and so by way of Balkh over the Pamir to Kashgar, and by way of Kotan and the Lob Nor into the Hwang-ho valley and on to Pekin. At Pekin was the Great Khan, and they were hospitably entertained.

Marco particularly pleased Kublai; he was young and clever, and it is clear he had mastered the Tartar language very thoroughly. He was given an official position and sent on several missions, chiefly in south-west China. The tale he had to tell of vast stretches of smiling and prosperous country, "all the way excellent hostelries for travellers," and "fine vineyards, fields and gardens," of "many abbeys" of Buddhist monks, of manufactures of "cloth of silk and gold and many fine taffetas," a "constant succession of cities and boroughs," and so on, first roused the incredulity and then fired the imagination of all Europe. He told of Burmah, and of its great armies with hundreds of elephants, and how these animals were defeated by the Mongol bowmen, and also of the Mongol conquest of Pegu. He told of Japan, and greatly exaggerated the amount of gold in that country. For three years Marco ruled the city of Yang-chow as governor, and he probably impressed the Chinese inhabitants as being very little more of a foreigner than any Tartar would have been. He may also have been sent on a mission to India. Chinese records mention a certain Polo attached to the imperial council in 1277, a very valuable confirmation of the general truth of the Polo story.

ANOTHER ANCIENT NEGRO BRONZE OF A EUROPEAN

(In the British Museum)

The publication of Marco Polo's travels produced a profound effect upon the European imagination. The European literature, and especially the European romance of the fifteenth century, echoes with the names in Marco Polo's story, with Cathay (North China) and Cambulac (Pekin) and the like.

Two centuries later, among the readers of the Travels of Marco Polo was a certain Genoese mariner, Christopher Columbus, who conceived the brilliant idea of sailing westward round the world to China. In Seville there is a copy of the Travels with marginal notes by Columbus. There were many reasons why the thought of a Genoese should be turned in this direction. Until its capture by the Turks in 1453 Constantinople had been an impartial trading mart between the Western world and the East, and the Genoese had traded there freely. But the "Latin" Venetians, the bitter rivals of the Genoese, had been the allies and helpers of the Turks against the Greeks, and with the coming of the Turks Constantinople turned an

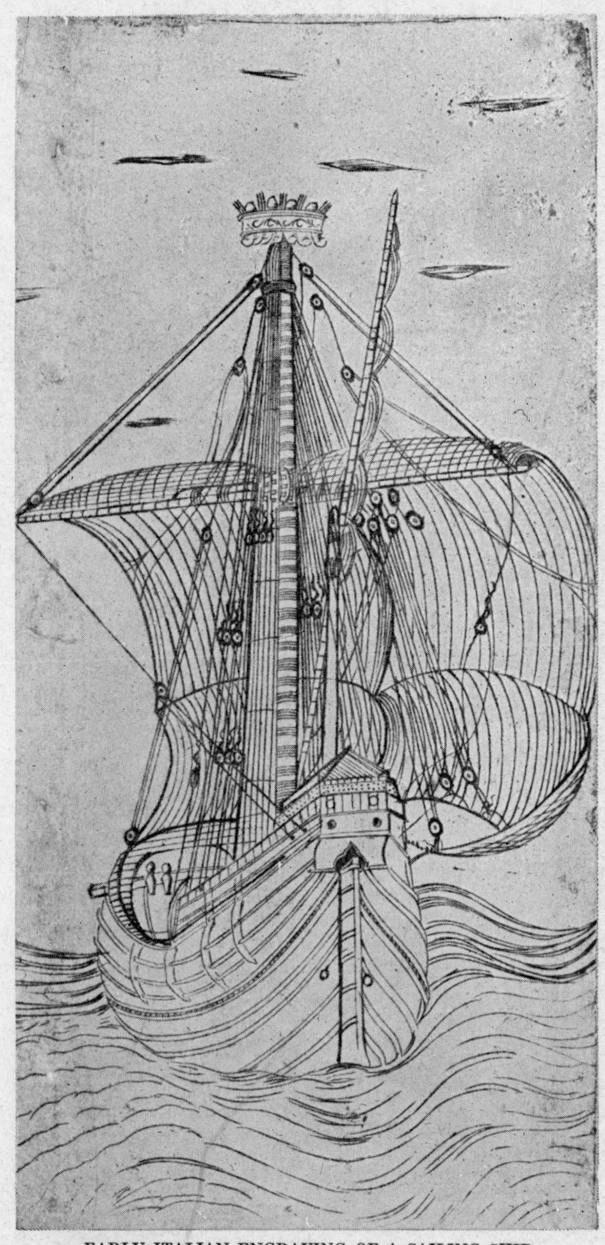

EARLY ITALIAN ENGRAVING OF A SAILING SHIP
(In the British Museum)

unfriendly face upon Genoese trade. The long forgotten discovery that the world was round had gradually resumed its sway over men's minds. The idea of going westward to China was therefore a fairly obvious one. It was encouraged by two things. The mariner's compass had now been invented and men were no longer left to the mercy of a fine night and the stars to determine the direction in which they were sailing, and the Normans, Catalonians and Genoese and Portuguese had already pushed out into the Atlantic as far as the Canary Isles, Madeira and the Azores.

Yet Columbus found many difficulties before he could get ships to put his idea to the test. He went from one European Court to another. Finally at Granada, just won from the Moors, he secured the patronage of Ferdinand and Isabella, and was able to set out across the unknown ocean in three small ships. After a voyage of two months and nine days he came to a land which he believed to be India, but which was really a new continent, whose distinct existence the old world had never hitherto suspected. He returned to Spain with gold, cotton, strange beasts and birds, and two wild-eyed painted Indians to be baptized. They were called Indians because, to the end of his days, he believed that this land he had found was India. Only in the course of several years did men begin to realize that the whole new continent of America was added to the world's resources.

The success of Columbus stimulated overseas enterprise enormously. In 1497 the Portuguese sailed round Africa to India, and in 1515 there were Portuguese ships in Java. In 1519 Magellan, a Portuguese sailor in Spanish employment, sailed out of Seville westward with five ships, of which one, the *Vittoria*, came back up the river to Seville in 1522, the first ship that had ever circumnavigated the world. Thirty-one men were aboard her, survivors of two-hundred-and-eighty who had started. Magellan himself had been killed in the Philippine Isles.

Printed paper books, a new realization of the round world as a thing altogether attainable, a new vision of strange lands, strange animals and plants, strange manners and customs, discoveries overseas and in the skies and in the ways and materials of life burst upon the European mind. The Greek classics, buried and forgotten for so long, were speedily being printed and studied, and were colouring men's thoughts with the dreams of Plato and the traditions of an age of republican freedom and dignity. The Roman dominion had first brought law and order to Western Europe, and the Latin Church had restored it; but under both Pagan and Catholic Rome curiosity and innovation were subordinate to and restrained by organization. The reign of the Latin mind was now drawing to an end. Between the thirteenth and the sixteenth century the European Aryans, thanks to the stimulating influence of Semite and Mongol and the rediscovery of the Greek classics, broke away from the Latin tradition and rose again to the intellectual and material leadership of mankind.

L

The Reformation of the Latin Church

THE Latin Church itself was enormously affected by this mental rebirth. It was dismembered; and even the portion that survived was extensively renewed.

We have told how nearly the church came to the autocratic leadership of all Christendom in the eleventh and twelfth centuries, and how in the fourteenth and fifteenth its power over men's minds and affairs declined. We have described how popular religious enthusiasm which had in earlier ages been its support and power was turned against it by its pride, persecutions and centralization, and how the insidious scepticism of Frederick II bore fruit in a growing insubordination of the princes. The Great Schism had reduced its religious and political prestige to negligible proportions. The forces of insurrection struck it now from both sides.

The teachings of the Englishman Wycliffe spread widely throughout Europe. In 1398 a learned Czech, John Huss, delivered a series of lectures upon Wycliffe's teachings in the university of Prague. This teaching spread rapidly beyond the educated class and aroused great popular enthusiasm. In 1414–18 a Council of the whole church was held at Constance to settle the Great Schism. Huss was invited to this Council under promise of a safe conduct from the emperor, seized, put on trial for heresy and burnt alive (1415). So far from tranquillizing the Bohemian people, this led to an insurrection of the Hussites in that country, the first of a series of religious wars that inaugurated the break-up of Latin Christendom. Against this insurrection Pope Martin V, the Pope specially elected at Constance as the head of a reunited Christendom, preached a Crusade.

Five Crusades in all were launched upon this sturdy little people and all of them failed. All the unemployed ruffianism of Europe was turned upon Bohemia in the fifteenth century, just as in the thirteenth it had been turned upon the Waldenses. But the Bohemian Czechs, unlike the Waldenses, believed in armed resistance. The Bohemian Crusade dissolved and streamed away from the battlefield at the sound of the Hussites' waggons and the distant chanting of their troops; it did not even wait to fight (battle of Domazlice, 1431). In 1436 an agreement was patched up with the Hussites by a new Council of the church at Basle in which many of the special objections to Latin practice were conceded.

AETHERNA IPSE SVAE MENTIS SIMVLACHRA LVTHERVS
EXPRIMIT· AT VVLTVS CERA LVCAE OCCIDVOS
·M·D·X·X·

PORTRAIT OF LUTHER

(From an early German engraving in the British Museum)

In the fifteenth century a great pestilence had produced much social disorganization throughout Europe. There had been extreme misery and discontent among the common people, and peasant risings against the landlords and the wealthy in England and France. After the Hussite Wars these peasant insurrections increased in gravity in Germany and took on a religious character. Printing came in as an influence upon this development. By the middle of the fifteenth century there were printers at work with movable type in Holland and the Rhineland. The art spread to Italy and England, where Caxton was printing in Westminster in 1477. The immediate consequence was a great increase and distribution of Bibles, and greatly increased facilities for widespread popular controversies. The European world became a world of readers, to an extent that had never happened to any community in the past. And this sudden irrigation of the general mind with clearer ideas and more accessible information occurred just at a time when the church was confused and divided and not in a position to defend itself effectively, and when many princes were looking for means to weaken its hold upon the vast wealth it claimed in their dominions.

In Germany the attack upon the church gathered round the personality of an ex-monk, Martin Luther (1483–1546), who appeared in Wittenberg in 1517 offering disputations against various orthodox doctrines and practices. At first he disputed in Latin in the fashion of the Schoolmen. Then he took up the new weapon of the printed word and scattered his views far and wide in German addressed to the ordinary people. An attempt was made to suppress him as Huss had been suppressed, but the printing press had changed conditions and he had too many open and secret friends among the German princes for this fate to overtake him.

For now in this age of multiplying ideas and weakened faith there were many rulers who saw their advantage in breaking the religious ties between their people and Rome. They sought to make themselves in person the heads of a more nationalized religion. England, Scotland, Sweden, Norway, Denmark, North Germany and Bohemia, one after another, separated themselves from the Roman Communion. They have remained separated ever since.

The various princes concerned cared very little for the moral and intellectual freedom of their subjects. They used the religious doubts and insurgence of their peoples to strengthen them against Rome, but they tried to keep a grip upon the popular movement as soon as that rupture was achieved and a national church set up under the control of the crown. But there has always been a curious vitality in the teaching of Jesus, a direct appeal to righteousness and a man's self-respect over every loyalty and every subordination, lay or ecclesiastical. None of these princely churches broke

A MAJOLICA DISH PAINTED IN COLOURS

An allegory of the Church triumphant over heretics and infidels. Italian (Urbino), dated 1543

(In the Victoria and Albert Museum)

off without also breaking off a number of fragmentary sects that would admit the intervention of neither prince nor Pope between a man and his God. In England and Scotland, for example, there was a number of sects who now held firmly to the Bible as their one guide in life and belief. They refused the disciplines of a state church. In England these dissentients were the Non-conformists, who played a very large part in the politics of that country in the seventeenth and eighteenth centuries. In England they carried their objection to a princely head to the church so far as to decapitate King Charles I (1649), and for eleven prosperous years England was a republic under Non-conformist rule.

The breaking away of this large section of Northern Europe from Latin Christendom is what is generally spoken of as the Reformation. But the shock and stress of these losses produced changes perhaps as profound in the Roman Church itself. The church was reorganized and a new spirit came into its life. One of the dominant figures in this revival was a young Spanish soldier, Inigo Lopez de Recalde, better known to the world as St. Ignatius of Loyola. After some romantic beginnings he became a priest (1538) and was permitted to found the Society of Jesus, a direct attempt to bring the generous and chivalrous traditions of military discipline into the service of religion. This Society of Jesus, the Jesuits, became one of the greatest teaching and missionary societies the world has ever seen. It carried Christianity to India, China and America. It arrested the rapid disintegration of the Roman Church. It raised the standard of education throughout the whole Catholic world; it raised the level of Catholic intelligence and quickened the Catholic conscience everywhere; it stimulated Protestant Europe to competitive educational efforts. The vigorous and aggressive Roman Catholic Church we know to-day is largely the product of this Jesuit revival.

LI

The Emperor Charles V

THE Holy Roman Empire came to a sort of climax in the reign of the Emperor Charles V. He was one of the most extraordinary monarchs that Europe has ever seen. For a time he had the air of being the greatest monarch since Charlemagne.

His greatness was not of his own making. It was largely the creation of his grandfather, the Emperor Maximilian I (1459–1519). Some families have fought, others have intrigued their way to world power; the Habsburgs married their way. Maximilian began his career with Austria, Styria, part of Alsace and other districts, the original Habsburg patrimony; he married — the lady's name scarcely matters to us — the Netherlands and Burgundy. Most of Burgundy slipped from him after his first wife's death, but the Netherlands he held. Then he tried unsuccessfully to marry Brittany. He became Emperor in succession to his father, Frederick III, in 1493, and married the duchy of Milan. Finally he married his son to the weak-minded daughter of Ferdinand and Isabella, the Ferdinand and Isabella of Columbus, who not only reigned over a freshly united Spain and over Sardinia and the kingdom of the two Sicilies, but over all America west of Brazil. So it was that this Charles V, his grandson, inherited most of the American continent and between a third and a half of what the Turks had left of Europe. He succeeded to the Netherlands in 1506. When his grandfather Ferdinand died in 1516, he became practically king of the Spanish dominions, his mother being imbecile; and his grandfather Maximilian dying in 1519, he was in 1520 elected Emperor at the still comparatively tender age of twenty.

He was a fair young man with a not very intelligent face, a thick upper lip and a long clumsy chin. He found himself in a world of young and vigorous personalities. It was an age of brilliant young monarchs. Francis I had succeeded to the French throne in 1515 at the age of twenty-one, Henry VIII had become King of England in 1509 at eighteen. It was the age of Baber in India (1526–1530) and Suleiman the Magnificent in Turkey (1520), both exceptionally capable monarchs, and the Pope Leo X (1513) was also a very distinguished Pope. The Pope and Francis I attempted to prevent the election of Charles as Emperor because they dreaded the concentration of so much power in the hands of one man. Both Francis I and Henry VIII offered themselves to the imperial electors. But there was now a long established tradition of Habsburg Emperors (since 1273), and some energetic bribery secured the election for Charles.

At first the young man was very much a magnificent puppet in the hands of his ministers. Then slowly he began to assert himself and take control. He began to realize something of the threatening complexities of his exalted position. It was a position as unsound as it was splendid.

From the very outset of his reign he was faced by the situation created by Luther's agitations in Germany. The Emperor had one reason for siding with the reformers in the opposition of the Pope to his election. But he had been brought up in Spain, that most Catholic of countries, and he decided against Luther. So he came into conflict with the Protestant princes and particularly the Elector of Saxony. He found himself in the presence of an opening rift that was to split the outworn fabric of Christendom into two contending camps. His attempts to close that rift were strenuous and honest and ineffective. There was an extensive peasant revolt in Germany which interwove with the general political and religious disturbance. And these internal troubles were complicated by attacks upon the Empire from east and west alike. On the west of Charles was his spirited rival, Francis I; to the east was the ever advancing Turk, who was now in Hungary, in alliance with Francis and clamouring for certain arrears of tribute from the Austrian dominions. Charles had the money and army of Spain at his disposal, but it was extremely difficult to get any effective support in money from Germany. His social and political troubles were complicated by financial distresses. He was forced to ruinous borrowing.

Photo: Anderson

THE CHARLES V PORTRAIT BY TITIAN

(*In the Gallery del Prado, Madrid*)

On the whole, Charles, in alliance with Henry VIII, was successful against Francis I and the Turk. Their chief battlefield was North Italy; the generalship was dull on both sides; their advances and retreats depended mainly on the arrival of reinforcements. The German army invaded France, failed to take Marseilles, fell back into Italy, lost Milan, and was besieged in Pavia. Francis I made a long and unsuccessful siege of Pavia, was caught by fresh German forces, defeated, wounded and taken prisoner. But thereupon the Pope and Henry VIII, still haunted by the fear of his attaining excessive power, turned against Charles. The German troops in Milan, under the Constable of Bourbon, being unpaid, forced rather than followed their commander into a raid upon Rome. They stormed the city and pillaged it (1527). The Pope took refuge in the Castle of St. Angelo while the looting and slaughter went on. He bought off the German troops at last by the payment of four hundred thousand ducats. Ten years of such confused fighting impoverished all Europe. At last the Emperor found himself triumphant in Italy. In 1530, he was crowned by the Pope — he was the last German Emperor to be so crowned — at Bologna.

Meanwhile the Turks were making great headway in Hungary. They had defeated and killed the king of Hungary in 1526, they held Buda-Pesth, and in 1529 Suleiman the Magnificent very nearly took Vienna. The Emperor was greatly concerned by these advances, and did his utmost to drive back the Turks, but he found the greatest difficulty in getting the German princes to unite even with this formidable enemy upon their very borders. Francis I remained implacable for a time, and there was a new French war; but in 1538 Charles won his rival over to a more friendly attitude after ravaging the south of France. Francis and Charles then formed an alliance against the Turk. But the Protestant princes, the German princes who were resolved to break away from Rome, had formed a league, the Schmalkaldic League, against the Emperor, and in the place of a great campaign to recover Hungary for Christendom Charles had to turn his mind to the gathering internal struggle in Germany. Of that struggle he saw only the opening war. It was a struggle, a sanguinary irrational bickering of princes, for ascendancy, now flaming into war and destruction, now sinking back to intrigues and diplomacies; it was a snake's sack of princely policies that was to go on writhing incurably right into the nineteenth century and to waste and desolate Central Europe again and again.

The Emperor never seems to have grasped the true forces at work in these gathering troubles. He was for his time and station an exceptionally worthy man, and he seems to have taken the religious dissensions that were tearing Europe into warring fragments as genuine theological differences. He gathered diets and councils in futile attempts at reconciliation. Formulæ and confessions were tried over. The student of German history must struggle with the details of the Religious Peace of Nuremberg, the settlement at the Diet of Ratisbon, the Interim of Augsburg, and the like. Here we do but mention them as details in the worried life of this culminating Emperor. As a matter of fact, hardly one of the multifarious princes and rulers in Europe seems to have been acting in good faith. The widespread religious trouble of the world, the desire of the common people for truth and social righteousness, the spreading knowledge of the time, all those things were merely counters in the imaginations of princely diplomacy. Henry VIII of England, who had begun his career with a book against heresy, and who had been rewarded by the Pope with the title of "Defender of the Faith," being anxious to divorce his first wife in favour of a young lady named Anne Boleyn, and wishing also to loot the vast wealth of the church in England, joined the company of Protestant princes in 1530. Sweden, Denmark and Norway had already gone over to the Protestant side.

The German religious war began in 1546, a few months after the death of Martin Luther. We need not trouble about the incidents of the campaign. The Protestant Saxon army was badly beaten at Lochau. By something very like a breach of faith Philip of Hesse, the Emperor's chief remaining antagonist, was caught and imprisoned, and the Turks were bought off by the promise of an annual tribute. In 1547, to the great relief of the Emperor, Francis I died. So by 1547 Charles got to a kind of settlement, and made his last efforts to effect peace where there was no peace. In 1552 all Germany was at war again, only a precipitate flight from Innsbruck saved Charles from capture, and in 1552, with the treaty of Passau, came another unstable equilibrium. . . .

Such is the brief outline of the politics of the Empire for thirty-two years. It is interesting to note how entirely the European mind was concentrated upon the struggle for European ascendancy. Neither Turks, French, English nor Germans had yet discovered any political interest in the great continent of America, nor any significance in the new sea routes to Asia. Great things were happening in America; Cortez with a mere handful of men had conquered the great Neolithic empire of Mexico for Spain, Pizarro had crossed the Isthmus of Panama (1530) and subjugated another wonder-land, Peru. But as yet these events meant no more to Europe than a useful and stimulating influx of silver to the Spanish treasury.

It was after the treaty of Passau that Charles began to display his distinctive originality of mind. He was now entirely bored and disillusioned by his imperial greatness. A sense of the intolerable futility of these European rivalries came upon him. He had never been of a very sound constitution, he was natually indolent and he was suffering greatly from gout. He abdicated. He made over all his sovereign rights in Germany to his brother Ferdinand, and Spain and the Netherlands he resigned to his son Philip. Then in a sort of magnificent dudgeon he retired to a monastery at Yuste, among the oak and chestnut forests in the hills to the north of the Tagus valley. There he died in 1558.

Much has been written in a sentimental vein of this retirement, this renunciation of the world by this tired majestic Titan, world-weary, seeking in an austere solitude his peace with God. But his retreat was neither solitary nor austere; he had with him nearly a hundred and fifty attendants; his establishment had all the splendour and indulgences without the fatigues of a court, and Philip II was a dutiful son to whom his father's advice was a command.

And if Charles had lost his living interest in the administration of European affairs, there were other motives of a more immediate sort to stir him. Says Prescott: "In the almost daily correspondence between Quixada, or Gaztelu, and the Secretary of State at Valladolid, there is scarcely a letter that does not turn more or less on the Emperor's eating or his illness. The one seems naturally to follow,

Photo: Alinari
INTERIOR OF ST. PETER'S, ROME, SHOWING THE HIGH ALTAR

like a running commentary, on the other. It is rare that such topics have formed the burden of communications with the department of state. It must have been no easy matter for the secretary to preserve his gravity in the perusal of despatches in which politics and gastronomy were so strangely mixed together. The courier from Valladolid to Lisbon was ordered to make a detour, so as to take Jarandilla in his route, and bring supplies to the royal table. On

Thursdays he was to bring fish to serve for the *jour maigre* that was to follow. The trout in the neighbourhood Charles thought too small, so others of a larger size were to be sent from Valladolid. Fish of every kind was to his taste, as, indeed, was anything that in its nature or habits at all approached to fish. Eels, frogs, oysters, occupied an important place in the royal bill of fare. Potted fish, especially anchovies, found great favour with him; and he regretted that he had not brought a better supply of these from the Low Countries. On an eel-pasty he particularly doted." . . .[1]

In 1554 Charles had obtained a bull from Pope Julius III granting him a dispensation from fasting, and allowing him to break his fast early in the morning even when he was to take the sacrament.

Eating and doctoring! it was a return to elemental things. He had never acquired the habit of reading, but he would be read aloud to at meals after the fashion of Charlemagne, and would make what one narrator describes as a "sweet and heavenly commentary." He also amused himself with mechanical toys, by listening to music or sermons, and by attending to the imperial business that still came drifting in to him. The death of the Empress, to whom he was greatly attached, had turned his mind towards religion, which in his case took a punctilious and ceremonial form; every Friday in Lent he scourged himself with the rest of the monks with such good will as to draw blood. These exercises and the gout released a bigotry in Charles that had hitherto been restrained by considerations of policy. The appearance of Protestant teaching close at hand in Valladolid roused him to fury. "Tell the grand inquisitor and his council from me to be at their posts, and to lay the axe at the root of the evil before it spreads further." . . . He expressed a doubt whether it would not be well, in so black an affair, to dispense with the ordinary course of justice, and to show no mercy; "lest the criminal, if pardoned, should have the opportunity of repeating his crime." He recommended, as an example, his own mode of proceeding in the Netherlands, "where all who remained obstinate in their errors were burned alive, and those who were admitted to penitence were beheaded."

And almost symbolical of his place and rôle in history was his

[1] Prescott's Appendix to Robertson's *History of Charles V.*

preoccupation with funerals. He seems to have had an intuition that something great was dead in Europe and sorely needed burial, that there was a need to write Finis, overdue. He not only attended every actual funeral that was celebrated at Yuste, but he had services conducted for the absent dead, he held a funeral service in memory of his wife on the anniversary of her death, and finally he celebrated his own obsequies.

"The chapel was hung with black, and the blaze of hundreds of wax-lights was scarcely sufficient to dispel the darkness. The brethren in their conventual dress, and all the Emperor's household clad in deep mourning, gathered round a huge catafalque, shrouded also in black, which had been raised in the centre of the chapel. The service for the burial of the dead was then performed; and, amidst the dismal wail of the monks, the prayers ascended for the departed spirit, that it might be received into the mansions of the blessed. The sorrowful attendants were melted to tears, as the image of their master's death was presented to their minds — or they were touched, it may be, with compassion by this pitiable display of weakness. Charles, muffled in a dark mantle, and bearing a lighted candle in his hand, mingled with his household, the spectator of his own obsequies; and the doleful ceremony was concluded by his placing the taper in the hands of the priest, in sign of his surrendering up his soul to the Almighty."

Within two months of this masquerade he was dead. And the brief greatness of the Holy Roman Empire died with him. His realm was already divided between his brother and his son. The Holy Roman Empire struggled on indeed to the days of Napoleon I but as an invalid and dying thing. To this day its unburied tradition still poisons the political air.

LII

The Age of Political Experiments; of Grand Monarchy and Parliaments and Republicanism in Europe

THE Latin Church was broken, the Holy Roman Empire was in extreme decay; the history of Europe from the opening of the sixteenth century onward is a story of peoples feeling their way darkly to some new method of government, better adapted to the new conditions that were arising. In the Ancient World, over long periods of time, there had been changes of dynasty and even changes of ruling race and language, but the form of government through monarch and temple remained fairly stable, and still more stable was the ordinary way of living. In this modern Europe since the sixteenth century the dynastic changes are unimportant, and the interest of history lies in the wide and increasing variety of experiments in political and social organization.

The political history of the world from the sixteenth century onward was, we have said, an effort, a largely unconscious effort, of mankind to adapt its political and social methods to certain new conditions that had now arisen. The effort to adapt was complicated by the fact that the conditions themselves were changing with a steadily increasing rapidity. The adaptation, mainly unconscious and almost always unwilling (for man in general hates voluntary change), has lagged more and more behind the alterations in conditions. From the sixteenth century onward the history of mankind is a story of political and social institutions becoming more and more plainly misfits, less comfortable and more vexatious, and of the slow reluctant realization of the need for a conscious and deliberate reconstruction of the whole scheme of human societies in the face of needs and possibilities new to all the former experiences of life.

What are these changes in the conditions of human life that have disorganized that balance of empire, priest, peasant and trader, with periodic refreshment by barbaric conquest, that has held human affairs in the Old World in a sort of working rhythm for more than a hundred centuries?

They are manifold and various, for human affairs are multitudinously complex; but the main changes seem all to turn upon one cause, namely the growth and extension of a knowledge of the nature of things, beginning first of all in small groups of intelligent people and spreading at first slowly, and in the last five hundred years very rapidly, to larger and larger proportions of the general population.

But there has also been a great change in human conditions due to a change in the spirit of human life. This change has gone on side by side with the increase and extension of knowledge, and is subtly connected with it. There has been an increasing disposition to treat a life based on the common and more elementary desires and gratifications as unsatisfactory, and to seek relationship with and service and participation in a larger life. This is the common characteristic of all the great religions that have spread throughout the world in the last twenty odd centuries, Buddhism, Christianity and Islam alike. They have had to do with the spirit of man in a way that the older religions did not have to do. They are forces quite different in their nature and effect from the old fetishistic blood-sacrifice religions of priest and temple that they have in part modified and in part replaced. They have gradually evolved a self-respect in the individual and a sense of participation and responsibility in the common concerns of mankind that did not exist among the populations of the earlier civilizations.

The first considerable change in the conditions of political and social life was the simplification and extended use of writing in the ancient civilizations which made larger empires and wider political understandings practicable and inevitable. The next movement forward came with the introduction of the horse, and later on of the camel as a means of transport, the use of wheeled vehicles, the extension of roads and the increased military efficiency due to the discovery of terrestrial iron. Then followed the profound economic disturbances due to the device of coined money and the change in the nature of debt, proprietorship and trade due to this convenient but dangerous convention. The empires grew in size and range, and men's ideas grew likewise to correspond with these things. Came the disappearance of local gods, the age of theocrasia, and the teaching of the great world religions. Came also the beginnings of reasoned and recorded history and geography, the first realization by man of his profound ignorance, and the first systematic search for knowledge.

For a time the scientific process which began so brilliantly in Greece and Alexandria was interrupted. The raids of the Teutonic barbarians, the westward drive of the Mongolian peoples, convulsive religious reconstruction and great pestilences put enormous strains upon political and social order. When civilization emerged again from this phase of conflict and confusion, slavery was no longer the basis of economic life; and the first paper-mills were preparing a new medium for collective information and co-operation in printed matter. Gradually at this point and that, the search for knowledge, the systematic scientific process, was resumed

And now from the sixteenth century onward, as an inevitable by-product of systematic thought, appeared a steadily increasing series of inventions and devices affecting the intercommunication and interaction of men with one another. They all tended towards wider range of action, greater mutual benefits or injuries, and increased co-operation, and they came faster and faster. Men's minds had not been prepared for anything of the sort, and until the great catastrophes at the beginning of the twentieth century quickened men's minds, the historian has very little to tell of any intelligently planned attempts to meet the new conditions this increasing flow of inventions was creating. The history of mankind for the last four centuries is rather like that of an imprisoned sleeper, stirring clumsily and uneasily while the prison that restrains and shelters him catches fire, not waking but incorporating the crackling and warmth of the fire with ancient and incongruous dreams, than like that of a man consciously awake to danger and opportunity.

Since history is the story not of individual lives but of communities, it is inevitable that the inventions that figure most in the historical record are inventions affecting communications. In the sixteenth century the chief new things that we have to note are the appearance of printed paper and the sea-worthy, ocean-going sailing ship using the new device of the mariner's compass. The former cheapened, spread, and revolutionized teaching, public information and discussion, and the fundamental operations of political activity. The latter made the round world one. But almost equally important was the increased utilization and improvement of guns and gunpowder which the Mongols had first brought westward in the thir-

CROMWELL DISSOLVES THE LONG PARLIAMENT AND SO BECOMES AUTOCRAT OF THE ENGLISH REPUBLIC

(From a contemporary satirical print in the British Museum)

teenth century. This destroyed the practical immunity of barons in their castles and of walled cities. Guns swept away feudalism. Constantinople fell to guns. Mexico and Peru fell before the terror of the Spanish guns.

The seventeenth century saw the development of systematic scientific publication, a less conspicuous but ultimately far more pregnant innovation. Conspicuous among the leaders in this great forward step was Sir Francis Bacon (1561–1626) afterwards Lord

Verulam, Lord Chancellor of England. He was the pupil and perhaps the mouthpiece of another Englishman, Dr. Gilbert, the experimental philosopher of Colchester (1540–1603). This second Bacon, like the first, preached observation and experiment, and he used the inspiring and fruitful form of a Utopian story, *The New Atlantis,* to express his dream of a great service of scientific research.

Presently arose the Royal Society of London, the Florentine Society, and later other national bodies for the encouragement of research and the publication and exchange of knowledge. These European scientific societies became fountains not only of countless inventions but also of a destructive criticism of the grotesque theological history of the world that had dominated and crippled human thought for many centuries.

Neither the seventeenth nor the eighteenth century witnessed any innovations so immediately revolutionary in human conditions as printed paper and the ocean-going ship, but there was a steady accumulation of knowledge and scientific energy that was to bear its full fruits in the nineteenth century. The exploration and mapping of the world went on. Tasmania, Australia, New Zealand appeared on the map. In Great Britain in the eighteenth century coal coke began to be used for metallurgical purposes, leading to a considerable cheapening of iron and to the possibility of casting and using it in larger pieces than had been possible before, when it had been smelted with wood charcoal. Modern machinery dawned.

Like the trees of the celestial city, science bears bud and flower and fruit at the same time and continuously. With the onset of the nineteenth century the real fruition of science — which indeed henceforth may never cease — began. First came steam and steel, the railway, the great liner, vast bridges and buildings, machinery of almost limitless power, the possibility of a bountiful satisfaction of every material human need, and then, still more wonderful, the hidden treasures of electrical science were opened to men. . . .

We have compared the political and social life of man from the sixteenth century onward to that of a sleeping prisoner who lies and dreams while his prison burns about him. In the sixteenth century the European mind was still going on with its Latin Imperial dream, its dream of a Holy Roman Empire, united under a Catholic Church. But just as some uncontrollable element in our composition will insist at times upon introducing into our dreams the most absurd and destructive comments, so thrust into this dream we find the sleeping face and craving stomach of the Emperor Charles V, while

THE COURT AT VERSAILLES
(From the print after Watteau in the British Museum)

Henry VIII of England and Luther tear the unity of Catholicism to shreds.

In the seventeenth and eighteenth centuries the dream turned to personal monarchy. The history of nearly all Europe during this period tells with variations the story of an attempt to consolidate a monarchy, to make it absolute and to extend its power over weaker adjacent regions, and of the steady resistance, first of the land-owners and then with the increase of foreign trade and home industry, of the growing trading and moneyed class, to the exaction and interference of the crown. There is no universal victory of either side; here it is the King who gets the upper hand while there it is the man of private property who beats the King. In one case we find a King becoming the sun and centre of his national world, while just over his borders a sturdy mercantile class maintains a republic. So wide a range of variation shows how entirely experimental, what local accidents, were all the various governments of this period.

A very common figure in these national dramas is the King's minister, often in the still Catholic countries a prelate, who stands behind the King, serves him and dominates him by his indispensable services.

Here in the limits set to us it is impossible to tell these various national dramas in detail. The trading folk of Holland went Protestant and republican, and cast off the rule of Philip II of Spain, the son of the Emperor Charles V. In England Henry VIII and his minister Wolsey, Queen Elizabeth and her minister Burleigh, prepared the foundations of an absolutism that was wrecked by the folly of James I and Charles I. Charles I was beheaded for treason to his people (1649), a new turn in the political thought of Europe. For a dozen years (until 1660) Britain was a republic; and the crown was an unstable power, much overshadowed by Parliament, until George III (1760–1820) made a strenuous and partly successful effort to restore its predominance. The King of France, on the other hand, was the most successful of all the European Kings in perfecting monarchy. Two great ministers, Richelieu (1585–1642) and Mazarin (1602–1661), built up the power of the crown in that country, and the process was aided by the long reign and very considerable abilities of King Louis XIV, "the Grand Monarque" (1643–1715).

Louis XIV was indeed the pattern King of Europe. He was, within his limitations, an exceptionally capable King; his ambition was stronger than his baser passions, and he guided his country towards bankruptcy through the complication of a spirited foreign policy with an elaborate dignity that still extorts our admiration. His immediate desire was to consolidate and extend France to the Rhine and Pyrenees, and to absorb the Spanish Netherlands; his remoter view saw the French Kings as the possible successors of Charlemagne in a recast Holy Roman Empire. He made bribery a state method almost more important than warfare. Charles II of

England was in his pay, and so were most of the Polish nobility, presently to be described. His money, or rather the money of the tax-paying classes in France, went everywhere. But his prevailing occupation was splendour. His great palace at Versailles with its salons, its corridors, its mirrors, its terraces and fountains and parks and prospects, was the envy and admiration of the world.

He provoked a universal imitation. Every king and princelet in Europe was building his own Versailles as much beyond his means as his subjects and credits would permit. Everywhere the nobility rebuilt or extended their chateaux to the new pattern. A great

THE SACK OF A VILLAGE DURING THE FRENCH REVOLUTION
(*From Callot's "Miseres de la Guerre"*)

industry of beautiful and elaborate fabrics and furnishings developed. The luxurious arts flourished everywhere; sculpture in alabaster, faience, gilt woodwork, metal work, stamped leather, much music, magnificent painting, beautiful printing and bindings, fine crockery, fine vintages. Amidst the mirrors and fine furniture went a strange race of "gentlemen" in tall powdered wigs, silks and laces, poised upon high red heels, supported by amazing canes; and still more wonderful "ladies," under towers of powdered hair and wearing vast expansions of silk and satin sustained on wire. Through it all postured the great Louis, the sun of his world, unaware of the meagre and sulky and bitter faces that watched him from those lower darknesses to which his sunshine did not penetrate.

The German people remained politically divided throughout this period of the monarchies and experimental governments, and a considerable number of ducal and princely courts aped the splendours of Versailles on varying scales. The Thirty Years' War (1618–48), a devastating scramble among the Germans, Swedes and Bohemians for fluctuating political advantages, sapped the energies of Germany for a century. A map must show the crazy patchwork in which this struggle ended, a map of Europe according to the peace of Westphalia (1648). One sees a tangle of principalities, dukedoms, free states and the like, some partly in and partly out of the Empire. Sweden's arm, the reader will note, reached far into Germany; and except for a few islands of territory within the imperial boundaries France was still far from the Rhine. Amidst this patchwork the

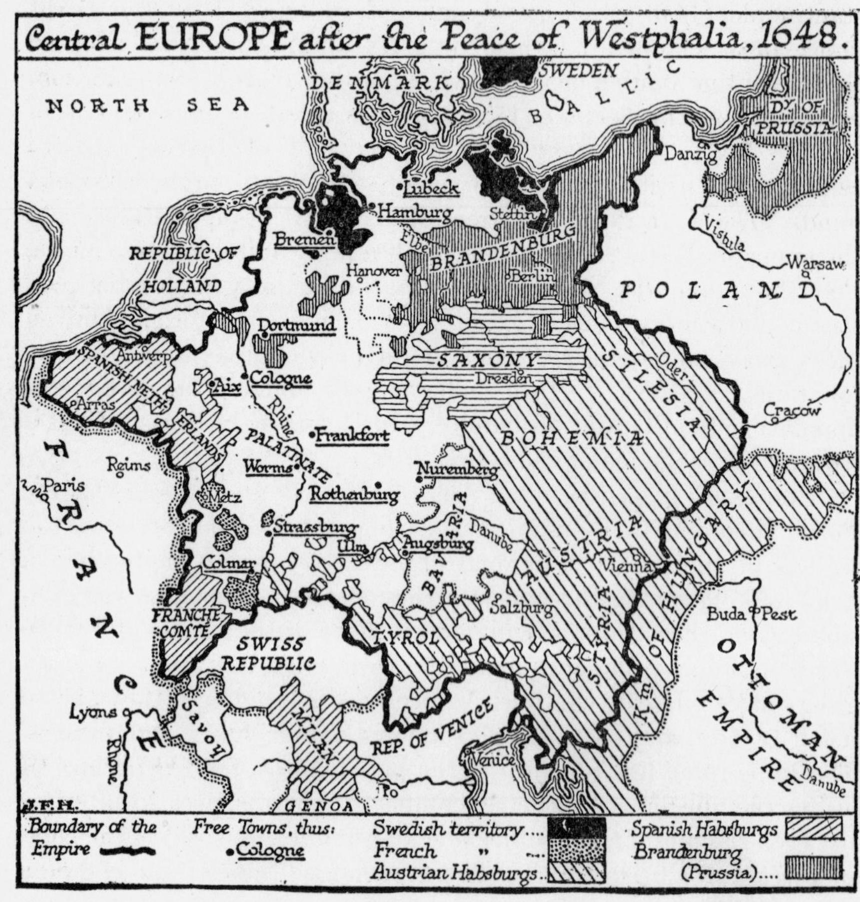

Central EUROPE after the Peace of Westphalia, 1648.

Kingdom of Prussia — it became a Kingdom in 1701 — rose steadily to prominence and sustained a series of successful wars. Frederick the Great of Prussia (1740–86) had his Versailles at Potsdam, where his court spoke French, read French literature and rivalled the culture of the French King.

In 1714 the Elector of Hanover became King of England, adding one more to the list of monarchies half in and half out of the empire.

The Austrian branch of the descendants of Charles V retained the title of Emperor; the Spanish branch retained Spain. But now there was also an Emperor of the East again. After the fall of Constantinople (1453), the grand duke of Moscow, Ivan the Great (1462–1505), claimed to be heir to the Byzantine throne and adopted the Byzantine double-headed eagle upon his arms. His grandson, Ivan IV, Ivan the Terrible (1533–1584), assumed the imperial title of Cæsar (Tsar). But only in the latter half of the seventeenth century did Russia cease to seem remote and Asiatic to the European mind. The Tsar Peter the Great (1682–1725) brought Russia into the arena of Western affairs. He built a new capital for his empire, Petersburg upon the Neva, that played the part of a window between Russia and Europe, and he set up his Versailles at Peterhof eighteen miles away, employing a French architect who gave him a terrace, fountains, cascades, picture gallery, park and all the recognized appointments of Grand Monarchy. In Russia as in Prussia French became the language of the court.

Unhappily placed between Austria, Prussia and Russia was the Polish kingdom, an ill-organized state of great landed proprietors too jealous of their own individual grandeur to permit more than a nominal kingship to the monarch they elected. Her fate was division among these three neighbours, in spite of the efforts of France to retain her as an independent ally. Switzerland at this time was a group of republican cantons; Venice was a republic; Italy like so much of Germany was divided among minor dukes and princes. The Pope ruled like a prince in the papal states, too fearful now of losing the allegiance of the remaining Catholic princes to interfere between them and their subjects or to remind the world of the commonweal of Christendom. There remained indeed no common political idea in Europe at all; Europe was given over altogether to division and diversity.

All these sovereign princes and republics carried on schemes of aggrandizement against each other. Each one of them pursued a "foreign policy" of aggression against its neighbours and of aggressive alliances. We Europeans still live to-day in the last phase of this age of the multifarious sovereign states, and still suffer from the hatreds, hostilities and suspicions it engendered. The history of this time becomes more and more manifestly "gossip," more and more unmeaning and wearisome to a modern intelligence. You are told of how this war was caused by this King's mistress, and how the jealousy of one minister for another caused that. A tittle-tattle of bribes and rivalries disgusts the intelligent student. The more permanently significant fact is that in spite of the obstruction of a score of frontiers, reading and thought still spread and increased and inventions multiplied. The eighteenth century saw the appearance of a literature profoundly sceptical and critical of the courts and policies of the time. In such a book as Voltaire's *Candide* we have the expression of an infinite weariness with the planless confusion of the European world.

LIII

The New Empires of the Europeans in Asia and Overseas

WHILE Central Europe thus remained divided and confused, the Western Europeans and particularly the Dutch, the Scandinavians, the Spanish, the Portuguese, the French and the British were extending the area of their struggles across the seas of all the world. The printing press had dissolved the political ideas of Europe into a vast and at first indeterminate fermentation, but that other great innovation, the ocean-going sailing ship, was inexorably extending the range of European experience to the furthermost limits of salt water.

The first overseas settlements of the Dutch and Northern Atlantic Europeans were not for colonization but for trade and mining. The Spaniards were first in the field; they claimed dominion over the whole of this new world of America. Very soon however the Portuguese asked for a share. The Pope — it was one of the last acts of Rome as mistress of the world — divided the new continent between these two first-comers, giving Portugal Brazil and everything else east of a line 370 leagues west of the Cape Verde islands, and all the rest to Spain (1494). The Portuguese at this time were also pushing overseas enterprise southward and eastward. In 1497 Vasco da Gama had sailed from Lisbon round the Cape to Zanzibar and then to Calicut in India. In 1515 there were Portuguese ships in Java and the Moluccas, and the Portuguese were setting up and fortifying trading stations round and about the coasts of the Indian Ocean. Mozambique, Goa, and two smaller possessions in India, Macao in China and a part of Timor are to this day Portuguese possessions.

The nations excluded from America by the papal settlement paid little heed to the rights of Spain and Portugal. The English, the Danes and Swedes, and presently the Dutch, were soon staking out claims in North America and the West Indies, and his Most Catholic Majesty of France heeded the papal settlement as little as any Protestant. The wars of Europe extended themselves to these claims and possessions.

In the long run the English were the most successful in this scramble for overseas possessions. The Danes and Swedes were too

EUROPEANS TIGER HUNTING IN INDIA

(From the engraving of the picture by Zoffany in the British Museum)

deeply entangled in the complicated affairs of Germany to sustain effective expeditions abroad. Sweden was wasted upon the German battlefields by a picturesque king, Gustavus Adolphus, the Protestant "Lion of the North." The Dutch were the heirs of such small settlements as Sweden made in America, and the Dutch were too near French aggressions to hold their own against the British. In the far East the chief rivals for empire were the British, Dutch and French, and in America the British, French and Spanish. The British had the supreme advantage of a water frontier, the "silver streak" of the English Channel, against Europe. The tradition of the Latin Empire entangled them least.

France has always thought too much in terms of Europe. Throughout the eighteenth century she was wasting her opportunities of expansion in West and East alike in order to dominate Spain, Italy and the German confusion. The religious and political dissensions of Britain in the seventeenth century had driven many of the English to seek a permanent home in America. They struck root and increased and multiplied, giving the British a great advantage in the American struggle. In 1756 and 1760 the French lost Canada to the British and their American colonists, and a few years later the British trading company found itself completely dominant over French, Dutch and Portuguese in the peninsula of India. The great Mongol Empire of Baber, Akbar and their successors had now far gone in decay, and the story of its practical capture by a London trading company, the British East India Company, is one of the most extraordinary episodes in the whole history of conquest.

This East India Company had been originally at the time of its incorporation under Queen Elizabeth no more than a company of sea adventurers. Step by step they had been forced to raise troops and arm their ships. And now this trading company, with its tradition of gain, found itself dealing not merely in spices and dyes and tea and jewels, but in the revenues and territories of princes

THE LAST EFFORT AND FALL OF TIPPOO SULTAN

(From the engraving of the picture by Singleton in the British Museum)

and the destinies of India. It had come to buy and sell, and it found itself achieving a tremendous piracy. There was no one to challenge its proceedings. Is it any wonder that its captains and commanders and officials, nay, even its clerks and common soldiers, came back to England loaded with spoils?

Men under such circumstances, with a great and wealthy land at their mercy, could not determine what they might or might not do. It was a strange land to them, with a strange sunlight; its brown people seemed a different race, outside their range of sympathy; its mysterious temples sustained fantastic standards of behaviour. Englishmen at home were perplexed when presently these generals and officials came back to make dark accusations against each other of extortions and cruelties. Upon Clive Parliament passed a vote of censure. He committed suicide in 1774. In 1788 Warren Hastings, a second great Indian administrator, was impeached and acquitted (1792). It was a strange and unprecedented situation in the world's history. The English Parliament found itself ruling over a London trading company, which in its turn was dominating an empire far greater and more populous than all the domains of the British crown. To the bulk of the English people India was a remote, fantastic, almost inaccessible land, to which adventurous poor young men went out, to return after many years very rich and very choleric old gentlemen. It was difficult for the English to conceive what the life of these countless brown millions in the eastern sunshine could be. Their imaginations declined the task. India remained romantically unreal. It was impossible for the English, therefore, to exert any effective supervision and control over the company's proceedings.

And while the Western European powers were thus fighting for these fantastic overseas empires upon every ocean in the world, two great land conquests were in progress in Asia. China had thrown off the Mongol yoke in 1360, and flourished under the great native dynasty of the Mings until 1644. Then the Manchus, another Mongol people, reconquered China and remained masters of China until 1912. Meanwhile Russia was pushing East and growing to greatness in the world's affairs. The rise of this great central power of the old world, which is neither altogether of the East nor altogether of the West, is one of the utmost importance to our human destiny. Its expansion is very largely due to the appearance of a Christian steppe people, the Cossacks, who formed a barrier between the feudal agriculture of Poland and Hungary to the west and the Tartar to the east. The Cossacks were the wild east of Europe, and in many ways not unlike the wild west of the United States in the middle nineteenth century. All who had made Russia too hot to hold them, criminals as well as the persecuted innocent, rebellious serfs, religious secretaries, thieves, vagabonds, murderers, sought asylum in the southern steppes and there made a fresh start and fought for life and freedom against Pole, Russian and Tartar alike. Doubtless fugitives from the Tartars to the east also contributed to the Cossack mixture. Slowly these border folk were incorporated in the Russian imperial service, much as the highland clans of Scotland were converted into regiments by the British government. New lands were offered them in Asia. They became a weapon against the dwindling power of the Mongolian nomads, first in Turkestan and then across Siberia as far as the Amur.

The decay of Mongol energy in the seventeenth and eighteenth centuries is very difficult to explain. Within two or three centuries from the days of Jengis and Timurlane Central Asia had relapsed from a period of world ascendancy to extreme political impotence. Changes of climate, unrecorded pestilences, infections of a malarial type, may have played their part in this recession — which may be only a temporary recession measured by the scale of universal history — of the Central Asian peoples. Some authorities think that the spread of Buddhist teaching from China also had a pacifying influence upon them. At any rate, by the sixteenth century the Mongol, Tartar and Turkish peoples were no longer pressing outward, but were being invaded, subjugated and pushed back both by Christian Russia in the west and by China in the east.

All through the seventeenth century the Cossacks were spreading eastward from European Russia, and settling wherever they found agricultural conditions. Cordons of forts and stations formed a moving frontier to these settlements to the south, where the Turkomans were still strong and active; to the north-east, however, Russia had no frontier until she reached right to the Pacific. . . .

LIV

The American War of Independence

THE third quarter of the eighteenth century thus saw the remarkable and unstable spectacle of a Europe divided against itself, and no longer with any unifying political or religious idea, yet through the immense stimulation of men's imaginations by the printed book, the printed map, and the opportunity of the new ocean-going shipping, able in a disorganized and contentious manner to dominate all the coasts of the world. It was a planless, incoherent ebullition of enterprise due to temporary and almost accidental advantages over the rest of mankind. By virtue of these advantages this new and still largely empty continent of America was peopled mainly from Western European sources, and South Africa and Australia and New Zealand marked down as prospective homes for a European population.

The motive that had sent Columbus to America and Vasco da Gama to India was the perennial first motive of all sailors since the beginning of things — trade. But while in the already populous and productive East the trade motive remained dominant, and the European settlements remained trading settlements from which the European inhabitants hoped to return home to spend their money, the Europeans in America, dealing with communities at a very much lower level of productive activity, found a new inducement for persistence in the search for gold and silver. Particularly did the mines of Spanish America yield silver. The Europeans had to go to America not simply as armed merchants but as prospectors, miners, searchers after natural products, and presently as planters. In the north they sought furs. Mines and plantations necessitated settlements. They obliged people to set up permanent overseas homes. Finally in some cases, as when the English Puritans went to New England in the early seventeenth century to escape religious persecution, when in the eighteenth Oglethorpe sent people from the English debtors' prisons to Georgia, and when in the end of the eighteenth the Dutch sent orphans to the Cape of Good Hope, the Europeans frankly crossed the seas to find new homes for good. In the nineteenth century, and especially after the coming of the steamship, the stream of European emigration to the new empty lands of America and Australia rose for some decades to the scale of a great migration.

So there grew up permanent overseas populations of Europeans, and the European culture was transplanted to much larger areas than those in which it had been developed. These new communities bringing a ready-made civilization with them to these new lands grew up, as it were, unplanned and unperceived; the statecraft of Europe did not foresee them, and was unprepared with any ideas about their treatment. The politicians and ministers of Europe continued to regard them as essentially expeditionary establishments, sources of revenue, "possessions" and "dependencies," long after their peoples had developed a keen sense of their separate social life. And also they continued to treat them as helplessly subject to the mother country long after the population had spread inland out of reach of any effectual punitive operations from the sea.

Because until right into the nineteenth century, it must be remembered, the link of all these overseas empires was the ocean-going sailing ship. On land the swiftest thing was still the horse, and the cohesion and unity of political systems on land was still limited by the limitations of horse communications.

Now at the end of the third quarter of the eighteenth century the northern two-thirds of North America was under the British crown. France had abandoned America. Except for Brazil, which was Portuguese, and one or two small islands and areas in French, British, Danish and Dutch hands, Florida, Louisiana, California and all America to the south was Spanish. It was the British colonies south of Maine and Lake Ontario that first demonstrated the inadequacy of the sailing ship to hold overseas populations together in one political system.

These British colonies were very miscellaneous in their origin and character. There were French, Swedish and Dutch settlements as well as British; there were British Catholics in Maryland and British ultra-Protestants in New England, and while the New Englanders farmed their own land and denounced slavery, the British in Virginia and the south were planters employing a swelling multitude of imported negro slaves. There was no natural common unity in such states. To get from one to the other might mean a coasting voyage hardly less tedious than the transatlantic crossing. But the union that diverse origin and natural conditions denied the British Americans was forced upon them by the selfishness and stupidity of the British government in London. They were taxed without any voice in the spending of the taxes; their trade was sacrificed to British interests; the highly profitable slave trade was maintained by the British government in spite of the opposition of the Virginians who — though quite willing to hold and use slaves — feared to be swamped by an ever-growing barbaric black population.

GEORGE WASHINGTON
(From a painting by Gilbert Stuart)

Britain at that time was lapsing towards an intenser form of monarchy, and the obstinate personality of George III (1760–1820) did much to force on a struggle between the home and the colonial governments.

The conflict was precipitated by legislation which favoured the London East India Company at the expense of the American shipper. Three cargoes of tea which were imported under the new conditions were thrown overboard in Boston harbour by a band of men disguised as Indians (1773).

THE BATTLE OF BUNKER HILL, NEAR BOSTON

(From the engraving of the picture by John Trumbull in the British Museum)

Fighting only began in 1775 when the British government attempted to arrest two of the American leaders at Lexington near Boston. The first shots were fired in Lexington by the British; the first fighting occurred at Concord.

So the American War of Independence began, though for more than a year the colonists showed themselves extremely unwilling to sever their links with the mother land. It was not until the middle of 1776 that the Congress of the insurgent states issued "The Declaration of Independence." George Washington, who like many of the leading colonists of the time had had a military training in the wars against the French, was made commander-in-chief. In 1777 a British general, General Burgoyne, in an attempt to reach New York from Canada, was defeated at Freemans Farm and obliged to surrender at Saratoga. In the same year the French and Spanish declared war upon Great Britain, greatly hampering her sea communications. A second British army under General Cornwallis was caught in the Yorktown peninsula in Virginia and obliged to capitulate in 1781. In 1783 peace was made in Paris, and the Thirteen Colonies from Maine to Georgia became a union of independent sovereign States. So the United States of America came into existence. Canada remained loyal to the British flag.

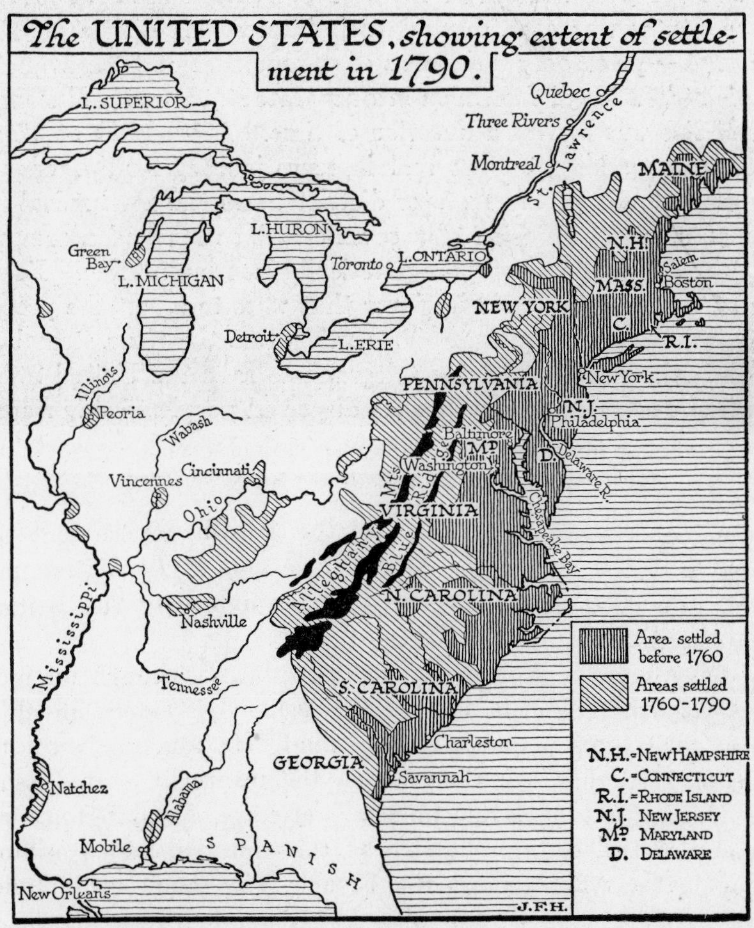

For four years these States had only a very feeble central government under certain Articles of Confederation, and they seemed destined to break up into separate independent communities. Their immediate separation was delayed by the hostility of the British and a certain aggressiveness on the part of the French which brought home to them the immediate dangers of division. A Constitution was drawn up and ratified in 1788 establishing a more efficient Federal government with a President holding very considerable powers, and the weak sense of national unity was invigorated by a second war with Britain in 1812. Nevertheless the area covered by the States was so wide and their interests so diverse at that time, that — given only the means of communication then available — a disintegration of the Union into separate states on the European scale of size was merely a question of time. Attendance at Washington meant a long, tedious and insecure journey for the senators and congressmen of the remoter districts, and the mechanical impediments to the diffusion of a common education and a common literature and intelligence were practically insurmountable. Forces were at work in the world however that were to arrest the process of differentiation altogether. Presently came the river steamboat and then the railway and the telegraph to save the United States from fragmentation, and weave its dispersed people together again into the first of great modern nations.

Twenty-two years later the Spanish colonies in America were to follow the example of the Thirteen and break their connection with Europe. But being more dispersed over the continent and separated by great mountainous chains and deserts and forests and by the Portuguese Empire of Brazil, they did not achieve a union among themselves. They became a constellation of republican states, very prone at first to wars among themselves and to revolutions.

Brazil followed a rather different line towards the inevitable separation. In 1807 the French armies under Napoleon had occupied the mother country of Portugal, and the monarchy had fled to Brazil. From that time on until they separated, Portugal was rather a dependency of Brazil than Brazil of Portugal. In 1822 Brazil declared itself a separate Empire under Pedro I, a son of the Portuguese King. But the new world has never been very favourable to monarchy. In 1889 the Emperor of Brazil was shipped off quietly to Europe, and the United States of Brazil fell into line with the rest of republican America.

LV

The French Revolution and the Restoration of Monarchy in France

BRITAIN had hardly lost the Thirteen Colonies in America before a profound social and political convulsion at the very heart of Grand Monarchy was to remind Europe still more vividly of the essentially temporary nature of the political arrangements of the world.

We have said that the French monarchy was the most successful of the personal monarchies in Europe. It was the envy and model of a multitude of competing and minor courts. But it flourished on a basis of injustice that led to its dramatic collapse. It was brilliant and aggressive, but it was wasteful of the life and substance of its common people. The clergy and nobility were protected from taxation by a system of exemption that threw the whole burden of the state upon the middle and lower classes. The peasants were ground down by taxation; the middle classes were dominated and humiliated by the nobility.

In 1787 this French monarchy found itself bankrupt and obliged to call representatives of the different classes of the realm into consultation upon the perplexities of defective income and excessive expenditure. In 1789 the States General, a gathering of the nobles, clergy and commons, roughly equivalent to the earlier form of the British Parliament, was called together at Versailles. It had not assembled since 1610. For all that time France had been an absolute monarchy. Now the people found a means of expressing their long fermenting discontent. Disputes immediately broke out between the three estates, due to the resolve of the Third Estate, the Commons, to control the Assembly. The Commons got the better of these disputes and the States General became a National Assembly, clearly resolved to keep the crown in order, as the British

Parliament kept the British crown in order. The king (Louis XVI) prepared for a struggle and brought up troops from the provinces. Whereupon Paris and France revolted.

The collapse of the absolute monarchy was very swift. The grim-looking prison of the Bastille was stormed by the people of Paris, and the insurrection spread rapidly throughout France. In the east and north-west provinces many chateaux belonging to the nobility were burnt by the peasants, their title-deeds carefully destroyed, and the owners murdered or driven away. In a month the ancient and decayed system of the aristocratic order had collapsed. Many of the leading princes and courtiers of the queen's party fled abroad. A provisional city government was set up in Paris and in most of the other large cities, and a new armed force, the National Guard, a force designed primarily and plainly to resist the forces of the crown, was brought into existence by these municipal bodies. The National Assembly found itself called upon to create a new political and social system for a new age.

It was a task that tried the powers of that gathering to the utmost. It made a great sweep of the chief injustices of the absolutist regime; it abolished tax exemptions, serfdom, aristocratic titles and privileges and sought to establish a constitutional monarchy in Paris. The king abandoned Versailles and its splendours and kept a diminished state in the palace of the Tuileries in Paris.

For two years it seemed that the National Assembly might struggle through to an effective modernized government. Much of its work was sound and still endures, if much was experimental and had to be undone. Much was ineffective. There was a clearing up of the penal code; torture, arbitrary imprisonment and persecutions for heresy were abolished. The ancient provinces of France, Normandy, Burgundy and the like gave place to eighty departments. Promotion to the highest ranks in the army was laid open to men of every class. An excellent and simple system of law courts was set up, but its value was much vitiated by having the judges appointed by popular election for short periods of time. This made the crowd a sort of final court of appeal, and the judges, like the members of the Assembly, were forced to play to the gallery. And the whole vast property of the church was seized and administered by the state; religious establishments not engaged in education or works of charity were broken up, and the salaries of the clergy made a charge upon the nation. This in itself was not a bad thing for the lower clergy in France, who were often scandalously underpaid in comparison with the richer dignitaries. But in addition the choice of priests and bishops was made elective, which struck at the very root idea of the Roman Church, which centred everything upon the Pope, and in which all authority is from above downward. Practically the National Assembly wanted at one blow to make the church in France Protestant, in organization if not in doctrine. Everywhere there were disputes and conflicts between the state priests created by the National Assembly and the recalcitrant (non-juring) priests who were loyal to Rome.

In 1791 the experiment of Constitutional monarchy in France was brought to an abrupt end by the action of the king and queen, working in concert with their aristocratic and monarchist friends abroad. Foreign armies gathered on the Eastern frontier and one night in June the king and queen and their children slipped away from the Tuileries and fled to join the foreigners and the aristocratic exiles. They were caught at Varennes and brought back to Paris, and all France flamed up into a passion of patriotic republicanism. A Republic was proclaimed, open war with Austria and Prussia ensued, and the king was tried and executed (January, 1793) on the model already set by England, for treason to his people.

And now followed a strange phase in the history of the French people. There arose a great flame of enthusiasm for France and the Republic. There was to be an end to compromise at home and abroad; at home royalists and every form of disloyalty were to be stamped out; abroad France was to be the protector and helper of all revolutionaries. All Europe, all the world, was to become Republican. The youth of France poured into the Republican armies; a new and wonderful song spread through the land, a song that still warms the blood like wine, the Marseillaise. Before that chant and the leaping columns of French bayonets and their enthusiastically served guns the foreign armies rolled back; before the end of 1792 the French armies had gone far beyond the utmost achievements of Louis XIV; everywhere they stood on foreign soil. They were in Brussels, they had overrun Savoy, they had raided to Mayence; they had seized the Scheldt from Holland. Then the French Government did an unwise thing. It had been exasperated by the expulsion of its representative from England upon the execution of Louis, and it declared war against England. It was an unwise thing to do, because the revolution which had given France a new enthusiastic infantry and a brilliant artillery

THE TRIAL OF LOUIS XVI
(From a print in the British Museum)

released from its aristocratic officers and many cramping conditions had destroyed the discipline of the navy, and the English were supreme upon the sea. And this provocation united all England against France, whereas there had been at first a very considerable liberal movement in Great Britain in sympathy with the revolution.

Of the fight that France made in the next few years against a European coalition we cannot tell in any detail. She drove the Austrians for ever out of Belgium, and made Holland a republic. The Dutch fleet, frozen in the Texel, surrendered to a handful of cavalry without firing its guns. For some time the French thrust towards Italy was hung up, and it was only in 1796 that a new general, Napoleon Bonaparte, led the ragged and hungry republican armies in triumph across Piedmont to Mantua and Verona. Says C. F. Atkinson,[1] "What astonished the Allies most of all was the number and the velocity of the Republicans. These improvised armies had in fact nothing to delay them. Tents were unprocurable for want of money, untransportable for want of the enormous number of wagons that would have been required, and also unnecessary, for the discomfort that would have caused wholesale desertion in professional armies was cheerfully borne by the men of 1793–94. Supplies for armies of then unheard-of size could not be carried in convoys, and the French soon became familiar with 'living on the country.' Thus 1793 saw the birth of the modern system of war — rapidity of movement, full development of national strength, bivouacs, requisitions and force as against cautious manœuvring, small professional armies, tents and full rations, and chicane. The first represented the decision-compelling spirit, the second the spirit of risking little to gain a little. . . ."

And while these ragged hosts of enthusiasts were chanting the Marseillaise and fighting for *la France*, manifestly never quite clear in their minds whether they were looting or liberating the countries into which they poured, the republican enthusiasm in Paris was spending itself in a far less glorious fashion. The revolution was now under the sway of a fanatical leader, Robespierre. This man is difficult to judge; he was a man of poor physique, naturally timid, and a prig. But he had that most necessary gift for power, faith. He set himself to save the Republic as he conceived it, and he imagined it could be saved by no other man than he. So that to keep in power was to save the Republic. The living spirit of the Republic, it seemed, had sprung from a slaughter of royalists and the execution of the king. There were insurrections; one in the west, in the district of La Vendée, where the people rose against the conscription and against the dispossession of the orthodox clergy, and were led by noblemen and priests; one in the south, where Lyons and Marseilles had risen and the royalists of Toulon

[1] In his article, "French Revolutionary Wars," in the Encyclopædia Britannica.

had admitted an English and Spanish garrison. To which there seemed no more effectual reply than to go on killing royalists.

The Revolutionary Tribunal went to work, and a steady slaughtering began. The invention of the guillotine was opportune to this mood. The queen was guillotined, most of Robespierre's antagonists were guillotined, atheists who argued that there was no Supreme Being were guillotined; day by day, week by week,

THE EXECUTION OF MARIE ANTOINETTE, QUEEN OF FRANCE, OCTOBER 16, 1793

(From a print in the British Museum)

this infernal new machine chopped off heads and more heads and more. The reign of Robespierre lived, it seemed, on blood; and needed more and more, as an opium-taker needs more and more opium.

Finally in the summer of 1794 Robespierre himself was overthrown and guillotined. He was succeeded by a Directory of five men which carried on the war of defence abroad and held France together at home for five years. Their reign formed a curious interlude in this history of violent changes. They took things as they found them. The propagandist zeal of the revolution carried the French armies into Holland, Belgium, Switzerland, south Germany and north Italy. Everywhere kings were expelled and republics set up. But such propagandist zeal as animated the Directorate did not prevent the looting of the treasures of the liberated peoples to relieve the financial embarrassment of the French Government. Their wars became less and less the holy wars of freedom, and more and more like the aggressive wars of the ancient regime. The last feature of Grand Monarchy that France was disposed to discard was her tradition of foreign policy. One discovers it still as vigorous under the Directorate as if there had been no revolution.

Unhappily for France and the world a man arose who embodied in its intensest form this national egotism of the French. He gave that country ten years of glory and the humiliation of a final defeat. This was that same Napoleon Bonaparte who had led the armies of the Directory to victory in Italy.

Throughout the five years of the Directorate he had been scheming and working for self-advancement. Gradually he clambered to supreme power. He was a man of severely limited understanding but of ruthless directness and great energy. He had begun life as an extremist of the school of Robespierre; he owed his first promotion to that side; but he had no real grasp of the new forces that were working in Europe. His utmost political imagination carried him to a belated and tawdry attempt to restore the Western Empire. He tried to destroy the remains of the old Holy Roman Empire, intending to replace it by a new one centring upon Paris. The Emperor in Vienna ceased to be the Holy Roman Emperor and became simply Emperor of Austria. Napoleon divorced his French wife in order to marry an Austrian princess.

He became practically monarch of France as First Consul in 1799, and he made himself Emperor of France in 1804 in direct imitation of Charlemagne. He was crowned by the Pope in Paris, taking the crown from the Pope and putting it upon his own head himself as Charlemagne had directed. His son was crowned King of Rome.

For some years Napoleon's reign was a career of victory. He

conquered most of Italy and Spain, defeated Prussia and Austria, and dominated all Europe west of Russia. But he never won the command of the sea from the British and his fleets sustained a conclusive defeat inflicted by the British Admiral Nelson at Trafalgar (1805). Spain rose against him in 1808 and a British army under Wellington thrust the French armies slowly northward out of the peninsula. In 1811 Napoleon came into conflict with the Tsar Alexander I, and in 1812 he invaded Russia with a great conglomerate army of 600,000 men, that was defeated and largely destroyed by the Russians and the Russian winter. Germany rose against him, Sweden turned against him. The French armies were beaten back and at Fontainebleau Napoleon abdicated (1814). He was exiled to Elba, returned to France for one last effort in 1815 and was defeated by the allied British, Belgians and Prussians at Waterloo. He died a British prisoner at St. Helena in 1821.

The forces released by the French revolution were wasted and finished. A great Congress of the victorious allies met at Vienna to restore as far as possible the state of affairs that the great storm had rent to pieces. For nearly forty years a sort of peace, a peace of exhausted effort, was maintained in Europe.

LVI

The Uneasy Peace in Europe that followed the Fall of Napoleon

TWO main causes prevented that period from being a complete social and international peace, and prepared the way for the cycle of wars between 1854 and 1871. The first of these was the tendency of the royal courts concerned, towards the restoration of unfair privilege and interference with freedom of thought and writing and teaching. The second was the impossible system of boundaries drawn by the diplomatists of Vienna.

The inherent disposition of monarchy to march back towards past conditions was first and most particularly manifest in Spain. Here even the Inquisition was restored. Across the Atlantic the Spanish colonies had followed the example of the United States and revolted against the European Great Power System, when Napoleon set his brother Joseph on the Spanish throne in 1810. The George Washington of South America was General Bolivar. Spain was unable to suppress this revolt, it dragged on much as the United States War of Independence had dragged on, and at last the suggestion was made by Austria, in accordance with the spirit of the Holy Alliance, that the European monarch should assist Spain in this struggle. This was opposed by Britain in Europe, but it was the prompt action of President Monroe of the United States in 1823 which conclusively warned off this projected monarchist restoration. He announced that the United States would regard any extension of the European system in the Western Hemisphere as a hostile act. Thus arose the Monroe Doctrine, the doctrine that there must be no extension of extra-American government in America, which has kept the Great Power system out of America for nearly a hundred years and permitted the new states of Spanish America to work out their destinies along their own lines.

But if Spanish monarchism lost its colonies, it could at least, under the protection of the Concert of Europe, do what it chose in Europe. A popular insurrection in Spain was crushed by a French army in 1823, with a mandate from a European congress, and simultaneously Austria suppressed a revolution in Naples.

In 1824 Louis XVIII died, and was succeeded by Charles X. Charles set himself to destroy the liberty of the press and universities, and to restore absolute government; the sum of a billion francs was voted to compensate the nobles for the chateau burnings and sequestrations of 1789. In 1830 Paris rose against this embodiment of the ancient regime, and replaced him by Louis Philippe, the son of that Philip, Duke of Orleans, who was executed during the Terror. The other continental monarchies, in face of the open approval of the revolution by Great Britain and a strong liberal ferment in Germany and Austria, did not interfere in this affair. After all, France was still a monarchy. This man Louis Philippe (1830–48) remained the constitutional King of France for eighteen years.

Such were the uneasy swayings of the peace of the Congress of Vienna, which were provoked by the reactionary proceedings of the monarchists. The stresses that arose from the unscientific boundaries planned by the diplomatists at Vienna gathered force more deliberately, but they were even more dangerous to the peace of mankind. It is extraordinarily inconvenient to administer together the affairs of peoples speaking different languages and so reading different literatures and having different general ideas, especially if those differences are exacerbated by religious disputes. Only some strong mutual interest, such as the common defensive needs of the Swiss mountaineers, can justify a close linking of peoples of dissimilar languages and faiths; and even in Switzerland there is the utmost local autonomy. When, as in Macedonia, populations are mixed in a patchwork of villages and districts, the cantonal system is imperatively needed. But if the reader will look at the map of Europe as the Congress of Vienna drew it, he will see that this gathering seems almost as if it had planned the maximum of local exasperation.

It destroyed the Dutch Republic, quite needlessly, it lumped together the Protestant Dutch with the French-speaking Catholics of the old Spanish (Austrian) Netherlands, and set up a kingdom of the Netherlands. It handed over not merely the old republic of Venice, but all of North Italy as far as Milan to the German-speaking Austrians. French-speaking Savoy it combined with pieces of Italy to restore the kingdom of Sardinia. Austria and Hungary, already a sufficiently explosive mixture of discordant nationalities, Germans, Hungarians, Czecho-Slovaks, Jugo-Slavs, Roumanians, and now Italians, was made still more impossible by confirming Austria's Polish acquisitions of 1772 and 1795. The Catholic and republican-spirited Polish people were chiefly given over to the less civilized rule of the Greek-orthodox Tsar, but important districts went to Protestant Prussia. The Tsar was also confirmed in his acquisition of the entirely alien Finns. The very dissimilar Norwegian and Swedish peoples were bound together under one king. Germany, the reader will see, was left in a particularly dangerous state of muddle. Prussia and Austria were both partly in and partly out of a German confederation, which included a multitude of minor states. The King of Denmark came into the German confederation by virtue of certain German-speaking possessions in Holstein. Luxembourg was included in the German confederation, though its ruler was also King of the Netherlands, and though many of its peoples talked French.

Here was a complete disregard of the fact that the people who talk German and base their ideas on German literature, the people who talk Italian and base their ideas on Italian literature, and the people who talk Polish and base their ideas on Polish literature, will all be far better off and most helpful and least obnoxious to the rest of mankind if they conduct their own affairs in their own idiom within the ring-fence of their own speech. Is it any wonder that one of the most popular songs in Germany during this period declared that wherever the German tongue was spoken, there was the German Fatherland!

In 1830 French-speaking Belgium, stirred up by the current revolution in France, revolted against its Dutch association in the kingdom of the Netherlands. The powers, terrified at the possibilities of a republic or of annexation to France, hurried in to pacify

PORTRAIT OF NAPOLEON (CORONATION)

(From a print in the British Museum)

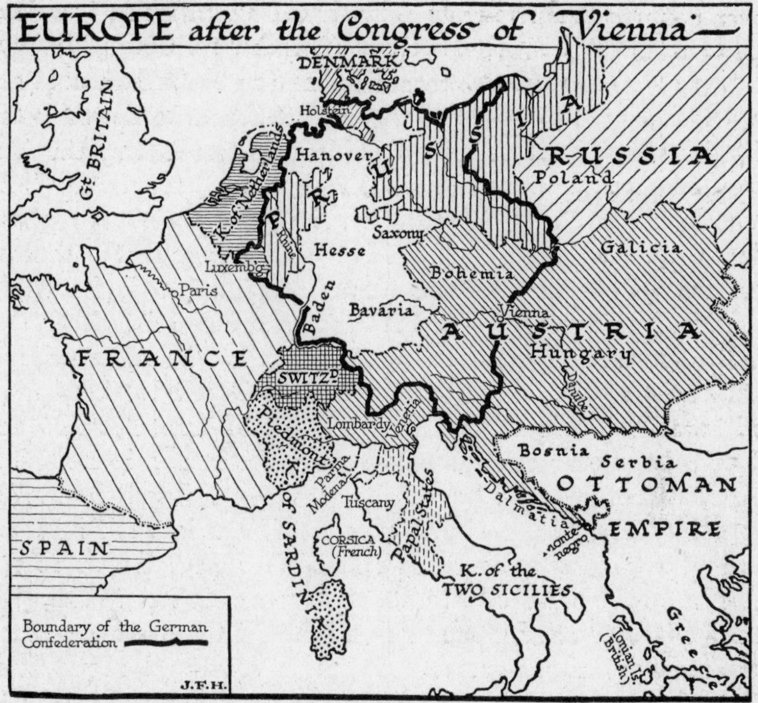

this situation, and gave the Belgians a monarch, Leopold I of Saxe-Coburg Gotha. There were also ineffectual revolts in Italy and Germany in 1830, and a much more serious one in Russian Poland. A republican government held out in Warsaw for a year against Nicholas I (who succeeded Alexander in 1825), and was then stamped out of existence with great violence and cruelty. The Polish language was banned, and the Greek Orthodox church was substituted for the Roman Catholic as the state religion. . . .

In 1821 there was an insurrection of the Greeks against the Turks. For six years they fought a desperate war, while the governments of Europe looked on. Liberal opinion protested against this inactivity; volunteers from every European country joined the insurgents, and at last Britain, France and Russia took joint action. The Turkish fleet was destroyed by the French and English at the battle of Navarino (1827), and the Tsar invaded Turkey. By the treaty of Adrianople (1829) Greece was declared free, but she was not permitted to resume her ancient republican traditions. A German king was found for Greece, one Prince Otto of Bavaria, and Christian governors were set up in the Danubian provinces (which are now Roumania) and Serbia (a part of the Jugo-Slav region). Much blood had still to run however before the Turk was altogether expelled from these lands.

LVII

The Development of Material Knowledge

THROUGHOUT the seventeenth and eighteenth centuries and the opening years of the nineteenth century, while these conflicts of the powers and princes were going on in Europe, and the patchwork of the treaty of Westphalia (1648) was changing kaleidoscopically into the patchwork of the treaty of Vienna (1815), and while the sailing ship was spreading European influence throughout the world, a steady growth of knowledge and a general clearing up of men's ideas about the world in which they lived was in progress in the European and Europeanized world.

It went on disconnected from political life, and producing throughout the seventeenth and eighteenth centuries no striking immediate results in political life. Nor was it affecting popular thought very profoundly during this period. These reactions were to come later, and only in their full force in the latter half of the nineteenth century. It was a process that went on chiefly in a small world of prosperous and independent-spirited people. Without what the English call the "private gentleman," the scientific process could not have begun in Greece, and could not have been renewed in Europe. The universities played a part but not a leading part in the philosophical and scientific thought of this period. Endowed learning is apt to be timid and conservative learning, lacking in initiative and resistent to innovation, unless it has the spur of contact with independent minds.

We have already noted the formation of the Royal Society in 1662 and its work in realizing the dream of Bacon's *New Atlantis*. Throughout the eighteenth century there was much clearing up of general ideas about matter and motion, much mathematical advance, a systematic development of the use of optical glass in microscope and telescope, a renewed energy in classificatory natural

EARLY ROLLING STOCK ON THE LIVERPOOL AND MANCHESTER RAILWAY IN THE FIRST DAYS OF THE RAILWAY

history, a great revival of anatomical science. The science of geology — foreshadowed by Aristotle and anticipated by Leonardo da Vinci (1452–1519) — began its great task of interpreting the Record of the Rocks.

The progress of physical science reacted upon metallurgy. Improved metallurgy, affording the possibility of a larger and bolder handling of masses of metal and other materials, reacted upon practical inventions. Machinery on a new scale and in a new abundance appeared to revolutionize industry.

In 1804 Trevithick adapted the Watt engine to transport and made the first locomotive. In 1825 the first railway, between Stockton and Darlington, was opened, and Stephenson's "Rocket," with a thirteen-ton train, got up to a speed of forty-four miles per hour. From 1830 onward railways multiplied. By the middle of the century a network of railways had spread all over Europe.

Here was a sudden change in what had long been a fixed condition of human life, the maximum rate of land transport. After the Russian disaster, Napoleon travelled from near Vilna to Paris in 312 hours. This was a journey of about 1,400 miles. He was travelling with every conceivable advantage, and he averaged

EARLY TRAVELLING ON THE LIVERPOOL AND MANCHESTER RAILWAY, 1833

under 5 miles an hour. An ordinary traveller could not have done this distance in twice the time. These were about the same maximum rates of travel as held good between Rome and Gaul in the first century A.D. Then suddenly came this tremendous change. The railways reduced this journey for any ordinary traveller to less than forty-eight hours. That is to say, they reduced the chief European distances to about a tenth of what they had been. They made it possible to carry out administrative work in areas ten times

THE STEAMBOAT: *CLERMONT*, 1807, U.S.A.

as great as any that had hitherto been workable under one administration. The full significance of that possibility in Europe still remains to be realized. Europe is still netted in boundaries drawn in the horse and road era. In America the effects were immediate. To the United States of America, sprawling westward, it meant the possibility of a continuous access to Washington, however far the frontier travelled across the continent. It meant unity, sustained on a scale that would otherwise have been impossible.

The steamboat was, if anything, a little ahead of the steam engine in its earlier phases. There was a steamboat, the *Charlotte Dundas*, on the Firth of Clyde Canal in 1802, and in 1807 an American named Fulton had a steamer, the *Clermont*, with British-built engines, upon the Hudson River above New York. The first steamship to put to sea was also an American, the *Phœnix*, which went from New York (Hoboken) to Philadelphia. So, too, was the first ship using steam (she also had sails) to cross the Atlantic, the *Savannah* (1819). All these were paddle-wheel boats and paddle-wheel boats are not adapted to work in heavy seas. The paddles smash too easily, and the boat is then disabled. The screw steamship followed rather slowly. Many difficulties had to be surmounted before the screw was a practicable thing. Not until the middle of the century did the tonnage of steamships upon the sea begin to overhaul that of sailing ships. After that the evolution in sea transport was rapid. For the first time men began to cross the seas and oceans with some certainty as to the date of their arrival. The transatlantic crossing, which had been an uncertain adventure of several weeks — which might stretch to months — was accelerated, until in 1910 it was brought down, in the case of the fastest boats, to under five days, with a practically notifiable hour of arrival.

Concurrently with the development of steam transport upon land and sea a new and striking addition to the facilities of human intercourse arose out of the investigations of Volta, Galvani and Faraday into various electrical phenomena. The electric telegraph came into existence in 1835. The first underseas cable was laid in 1851 between France and England. In a few years the telegraph system had spread over the civilized world, and news which had hitherto travelled slowly from point to point became practically simultaneous throughout the earth.

These things, the steam railway and the electric telegraph, were to the popular imagination of the middle nineteenth century the most striking and revolutionary of inventions, but they were only the most conspicuous and clumsy first fruits of a far more extensive process. Technical knowledge and skill were developing with an extraordinary rapidity, and to an extraordinary extent measured by the progress of any previous age. Far less conspicuous at first in everyday life, but finally far more important, was the extension of man's power over various structural materials. Before the middle of the eighteenth century iron was reduced from its ores by means of wood charcoal, was handled in small pieces, and hammered and wrought into shape. It was material for a craftsman. Quality and treatment were enormously dependent upon the experience and sagacity of the individual iron-worker. The largest masses of iron that could be dealt with under those conditions amounted at most (in the sixteenth century) to two or three tons. (There was a very definite upward limit, therefore, to the size of cannon.) The blast-furnace rose in the eighteenth century and developed with the use of coke. Not before the eighteenth century do we find rolled sheet iron (1728) and rolled rods and bars (1783). Nasmyth's steam hammer came as late as 1838.

The ancient world, because of its metallurgical inferiority, could not use steam. The steam engine, even the primitive pumping engine, could not develop before sheet iron was available. The early engines seem to the modern eye very pitiful and clumsy bits of ironmongery, but they were the utmost that the metallurgical science of the time could do. As late as 1856 came the Bessemer process, and presently (1864) the open-hearth process, in which steel and every sort of iron could be melted, purified and cast in a manner and upon a scale hitherto unheard of. To-day in the electric furnace one may see tons of incandescent steel swirling about like boiling milk in a saucepan. Nothing in the previous practical advances of mankind is comparable in its consequences to the complete mastery over enormous masses of steel and iron and over their texture and quality which man has now achieved. The railways and early engines of all sorts were the mere first triumphs of the new metallurgical methods. Presently came ships of iron and steel, vast bridges, and a new way of building with steel upon a gigantic scale. Men realized too late that they had planned their railways with far too timid a gauge, that they could have organized their travelling with far more steadiness and comfort upon a much bigger scale.

Before the nineteenth century there were no ships in the world much over 2,000 tons burthen; now there is nothing wonderful about a 50,000-ton liner. There are people who sneer at this kind of progress as being a progress in "mere size," but that sort of sneering merely marks the intellectual limitations of those who indulge in it.

The great ship or the steel-frame building is not, as they imagine, a magnified version of the small ship or building of the past; it is a thing different in kind, more lightly and strongly built, of finer and stronger materials; instead of being a thing of precedent and rule-of-thumb, it is a thing of subtle and intricate calculation. In the old house or ship, matter was dominant — the material and its needs had to be slavishly obeyed; in the new, matter had been captured, changed, coerced. Think of the coal and iron and sand dragged out of the banks and pits, wrenched, wrought, molten and cast, to be flung at last, a slender glittering pinnacle of steel and glass, six hundred feet above the crowded city!

We have given these particulars of the advance in man's knowledge of the metallurgy of steel and its results by way of illustration. A parallel story could be told of the metallurgy of copper and tin, and of a multitude of metals, nickel and aluminium to name but two, unknown before the nineteenth century dawned. It is in this great and growing mastery over substances, over different sorts of glass, over rocks and plasters and the like, over colours and textures, that the main triumphs of the mechanical revolution have thus far been achieved. Yet we are still in the stage of the first fruits in the matter. We have the power, but we have still to learn how to use our power. Many of the first employments of these gifts of science have been vulgar, tawdry, stupid or horrible. The artist and the adaptor have still hardly begun to work with the endless variety of substances now at their disposal.

Parallel with this extension of mechanical possibilities the new science of electricity grew up. It was only in the eighties of the nineteenth century that this body of enquiry began to yield results to impress the vulgar mind. Then suddenly came electric light and electric traction, and the transmutation of forces, the possibility of sending *power*, that could be changed into mechanical motion or light or heat as one chose, along a copper wire, as water is sent along a pipe, began to come through to the ideas of ordinary people. . . .

The British and French were at first the leading peoples in this great proliferation of knowledge; but presently the Germans, who had learnt humility under Napoleon, showed such zeal and pertinacity in scientific enquiry as to overhaul these leaders. British

EIGHTEENTH CENTURY SPINNING WHEEL
(In the Ipswich Museum)

science was largely the creation of Englishmen and Scotchmen working outside the ordinary centres of erudition.

The universities of Britain were at this time in a state of educational retrogression, largely given over to a pedantic conning of the Latin and Greek classics. French education, too, was dominated by the classical tradition of the Jesuit schools, and consequently it was not difficult for the Germans to organize a body of investigators, small indeed in relation to the possibilities of the case, but large in proportion to the little band of British and French inventors and experimentalists. And though this work of research and experiment was making Britain and France the most rich and

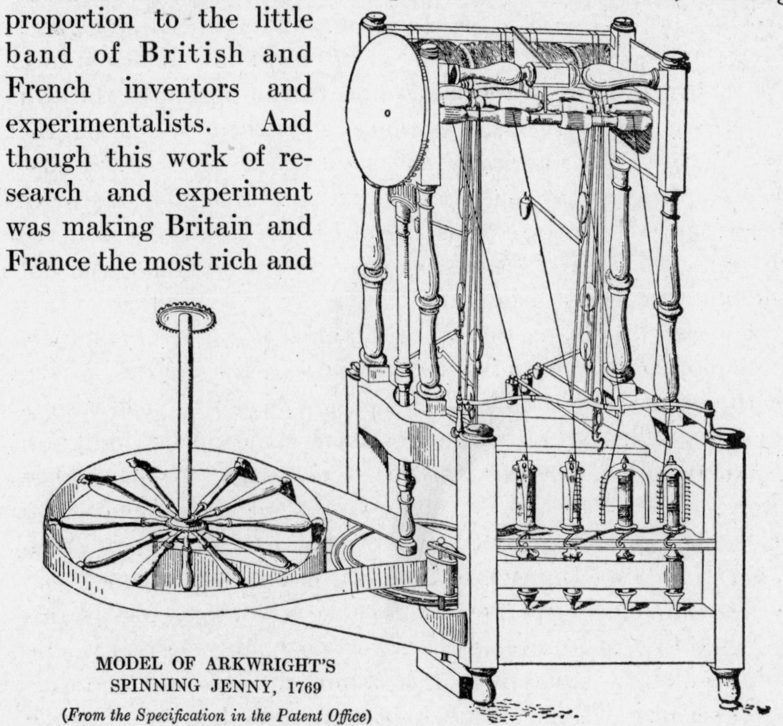

MODEL OF ARKWRIGHT'S SPINNING JENNY, 1769
(From the Specification in the Patent Office)

powerful countries in the world, it was not making scientific and inventive men rich and powerful. There is a necessary unworldliness about a sincere scientific man; he is too preoccupied with his research to plan and scheme how to make money out of it. The economic exploitation of his discoveries falls very easily and naturally, therefore, into the hands of a more acquisitive type; and so we find that the crops of rich men which every fresh phase of scientific and technical progress has produced in Great Britain, though they have not displayed quite the same passionate desire to insult and kill the goose that laid the national golden eggs as the scholastic and clerical professions, have been quite content to let that profitable creature starve. Inventors and discoverers came by nature, they thought, for cleverer people to profit by.

In this matter the Germans were a little wiser. The German "learned" did not display the same vehement hatred of the new learning. They permitted its development. The German business man and manufacturer again had not quite the same contempt for the man of science as had his British competitor. Knowledge, these Germans believed, might be a cultivated crop, responsive to fertilizers. They did concede, therefore, a certain amount of opportunity to the scientific mind; their public expenditure on scientific work was relatively greater, and this expenditure was abundantly rewarded. By the latter half of the nineteenth century the German scientific worker had made German a necessary language for every science student who wished to keep abreast with the latest work in his department, and in certain branches, and particularly in chemistry, Germany acquired a very great superiority over her western neighbours. The scientific effort of the sixties and seventies in Germany began to tell after the eighties, and the German gained steadily upon Britain and France in technical and industrial prosperity.

A fresh phase in the history of invention opened when in the eighties a new type of engine came into use, an engine in which the expansive force of an explosive mixture replaced the expansive force of steam. The light, highly efficient engines that were thus made possible were applied to the automobile, and developed at last to reach such a pitch of lightness and efficiency as to render flight—

AN EARLY WEAVING MACHINE
(From an engraving by W. Hincks in the British Museum)

long known to be possible — a practical achievement. A successful flying machine — but not a machine large enough to take up a human body — was made by Professor Langley of the Smithsonian Institute of Washington as early as 1897. By 1909 the aeroplane was available for human locomotion. There had seemed to be a pause in the increase of human speed with the perfection of railways and automobile road traction, but with the flying machine came fresh reductions in the effective distance between one point of the earth's surface and another. In the eighteenth century the distance from London to Edinburgh was an eight days' journey; in 1918 the British Civil Air Transport Commission reported that the journey from London to Melbourne, halfway round the earth, would probably in a few years' time be accomplished in that same period of eight days.

Too much stress must not be laid upon these striking reductions in the time distances of one place from another. They are merely one aspect of a much profounder and more momentous enlargement of human possibility. The science of agriculture and agricultural chemistry, for instance, made quite parallel advances during the nineteenth century. Men learnt so to fertilize the soil as to produce quadruple and quintuple the crops got from the same area in the seventeenth century. There was a still more extraordinary advance in medical science; the average duration of life rose, the daily efficiency increased, the waste of life through ill-health diminished.

Now here altogether we have such a change in human life as to constitute a fresh phase of history. In a little more than a century this mechanical revolution has been brought about. In that time man made a stride in the material conditions of his life vaster than he had done during the whole long interval between the palæolithic stage and the age of cultivation, or between the days of Pepi in Egypt and those of George III. A new gigantic material framework for human affairs has come into existence. Clearly it demands great readjustments of our social, economical and political methods. But these readjustments have necessarily waited upon the development of the mechanical revolution, and they are still only in their opening stage to-day.

LVIII

The Industrial Revolution

THERE is a tendency in many histories to confuse together what we have here called the *mechanical revolution*, which was an entirely new thing in human experience arising out of the development of organized science, a new step like the invention of agriculture or the discovery of metals, with something else, quite different in its origins, something for which there was already an historical precedent, the social and financial development which is called the *industrial revolution*. The two processes were going on together, they were constantly reacting upon each other, but they were in root and essence different. There would have been an industrial revolution of sorts if there had been no coal, no steam, no machinery; but in that case it would probably have followed far more closely upon the lines of the social and financial developments of the later years of the Roman Republic. It would have repeated the story of dispossessed free cultivators, gang labour, great estates, great financial fortunes, and a socially destructive financial process. Even the factory method came before power and machinery. Factories were the product not of machinery, but of the "division of labour." Drilled and sweated workers were making such things as millinery cardboard boxes and furniture, and colouring maps and book illustrations and so forth, before even water-wheels had been used for industrial purposes. There were factories in Rome in the days of Augustus. New books, for instance, were dictated to rows of copyists in the factories of the book-sellers. The attentive student of Defoe and of the political pamphlets of Fielding will realize that the idea of herding poor people into establishments to work collectively for their living was already current in Britain before the close of the seventeenth century. There are intimations of it even as early as More's *Utopia* (1516). It was a social and not a mechanical development.

Up to past the middle of the eighteenth century the social and economic history of western Europe was in fact retreading the path along which the Roman state had gone in the last three centuries B.C. But the political disunions of Europe, the political convulsions against monarchy, the recalcitrance of the common folk and perhaps also the greater accessibility of the western European intelligence to mechanical ideas and inventions, turned the process into quite novel directions. Ideas of human solidarity, thanks to Christianity, were far more widely diffused in the newer European world, political power was not so concentrated, and the man of energy anxious to get rich turned his mind, therefore, very willingly from the ideas of the slave and of gang labour to the idea of mechanical power and the machine.

The mechanical revolution, the process of mechanical invention and discovery, was a new thing in human experience and it went on regardless of the social, political, economic and industrial consequences it might produce. The industrial revolution, on the other hand, like most other human affairs, was and is more and more profoundly changed and deflected by the constant variation in human conditions caused by the mechanical revolution. And the essential difference between the amassing of riches, the extinction of small farmers and small business men, and the phase of big finance in the latter centuries of the Roman Republic on the one hand, and the very similar concentration of capital in the eighteenth and nineteenth centuries on the other, lies in the profound difference in the character of labour that the mechanical revolution was bringing about. The power of the old world was human power; everything depended ultimately upon the driving power of human muscle, the muscle of ignorant and subjugated men. A little animal muscle, supplied by draft oxen, horse traction and the like, contributed. Where a weight had to be lifted, men lifted it; where a rock had to be quarried, men chipped it out; where a field had to be ploughed, men and oxen ploughed it; the Roman equivalent of the steamship was the galley with its bank of sweating rowers. A vast proportion of mankind in the early civilizations were employed in purely mechanical drudgery. At its onset, power-driven machinery did not seem to promise any release from such unintelligent toil. Great gangs

INCIDENT IN THE DAYS OF THE SLAVE TRADE
(From a print after Morland in the British Museum)

of men were employed in excavating canals, in making railway cuttings and embankments, and the like. The number of miners increased enormously. But the extension of facilities and the output of commodities increased much more. And as the nineteenth century went on, the plain logic of the new situation asserted itself more clearly. Human beings were no longer wanted as a source of mere indiscriminated power. What could be done mechanically by a human being could be done faster and better by a machine. The human being was needed now only where choice and intelligence had to be exercised. Human beings were wanted only as human beings. The *drudge*, on whom all the previous civilizations had rested, the creature of mere obedience, the man whose brains were superfluous, had become unnecessary to the welfare of mankind.

This was as true of such ancient industries as agriculture and mining as it was of the newest metallurgical processes. For ploughing, sowing and harvesting, swift machines came forward to do the work of scores of men. The Roman civilization was built upon cheap and degraded human beings; modern civilization is being rebuilt upon cheap mechanical power. For a hundred years power has been getting cheaper and labour dearer. If for a generation or so machinery has had to wait its turn in the mine, it is simply because for a time men were cheaper than machinery.

Now here was a change-over of quite primary importance in human affairs. The chief solicitude of the rich and of the ruler in the old civilization had been to keep up a supply of drudges. As the nineteenth century went on, it became more and more plain to the intelligent directive people that the common man had now to be something better than a drudge. He had to be educated — if only to secure "industrial efficiency." He had to understand what he was about. From the days of the first Christian propaganda, popular education had been smouldering in Europe, just as it had smouldered in Asia wherever Islam has set its foot, because of the necessity of making the believer understand a little of the belief by which he is

EARLY FACTORY, IN COLEBROOKDALE

(From a print in the British Museum)

saved, and of enabling him to read a little in the sacred books by which his belief is conveyed. Christian controversies, with their competition for adherents, ploughed the ground for the harvest of popular education. In England, for instance, by the thirties and forties of the nineteenth century, the disputes of the sects and the necessity of catching adherents young had produced a series of competing educational organizations for children, the church "National" schools, the dissenting "British" schools, and even Roman Catholic elementary schools. The second half of the nineteenth century was a period of rapid advance in popular education throughout all the Westernized world. There was no parallel advance in the education of the upper classes — some advance, no doubt, but nothing to correspond — and so the great gulf that had divided that world hitherto into the readers and the non-reading mass became little more than a slightly perceptible difference in educational level. At the back of this process was the mechanical revolution, apparently regardless of social conditions, but really insisting inexorably upon the complete abolition of a totally illiterate class throughout the world.

The economic revolution of the Roman Republic had never been clearly apprehended by the common people of Rome. The ordinary Roman citizen never saw the changes through which he lived, clearly and comprehensively as we see them. But the industrial revolution, as it went on towards the end of the nineteenth century, was more and more distinctly *seen* as one whole process by the common people it was affecting, because presently they could read and discuss and communicate, and because they went about and saw things as no commonalty had ever done before.

LIX

The Development of Modern Political and Social Ideas

THE institutions and customs and political ideas of the ancient civilizations grew up slowly, age by age, no man designing and no man foreseeing. It was only in that great century of human adolescence, the sixth century B.C., that men began to think clearly about their relations to one another, and first to question and first propose to alter and rearrange the established beliefs and laws and methods of human government.

We have told of the glorious intellectual dawn of Greece and Alexandria, and how presently the collapse of the slave-holding civilizations and the clouds of religious intolerance and absolutist government darkened the promise of that beginning. The light of fearless thinking did not break through the European obscurity again effectually until the fifteenth and sixteenth centuries. We have tried to show something of the share of the great winds of Arab curiosity and Mongol conquest in this gradual clearing of the mental skies of Europe. And at first it was chiefly material knowledge that increased. The first fruits of the recovered manhood of the race were material achievements and material power. The science of human relationship, of individual and social psychology, of education and of economics, are not only more subtle and intricate in themselves but also bound up inextricably with much emotional matter. The advances made in them have been slower and made against greater opposition. Men will listen dispassionately to the most diverse suggestions about stars or molecules, but ideas about our ways of life touch and reflect upon everyone about us.

And just as in Greece the bold speculations of Plato came before Aristotle's hard search for fact, so in Europe the first political enquiries of the new phase were put in the form of "Utopian" stories, directly imitated from Plato's *Republic* and his *Laws*. Sir Thomas

More's *Utopia* is a curious imitation of Plato that bore fruit in a new English poor law. The Neapolitan Campanella's *City of the Sun* was more fantastic and less fruitful.

By the end of the seventeenth century we find a considerable and growing literature of political and social science was being produced. Among the pioneers in this discussion was John Locke, the son of an English republican, an Oxford scholar who first directed his attention to chemistry and medicine. His treatises on government, toleration and education show a mind fully awake to the possibilities of social reconstruction. Parallel with and a little later than John Locke in England, Montesquieu (1689–1755) in France subjected social, political and religious institutions to a searching and fundamental analysis. He stripped the magical prestige from the absolutist monarchy in France. He shares with Locke the credit for clearing away many of the false ideas that had hitherto prevented deliberate and conscious attempts to reconstruct human society.

The generation that followed him in the middle and later decades of the eighteenth century was boldly speculative upon the moral and intellectual clearings he had made. A group of brilliant writers, the "Encyclopædists," mostly rebel spirits from the excellent schools of the Jesuits, set themselves to scheme out a new world (1766). Side by side with the Encyclopædists were the Economists or Physiocrats, who were making bold and crude enquiries into the production and distribution of food and goods. Morelly, the author of the *Code de la Nature,* denounced the institution of private property and proposed a communistic organization of society. He was the precursor of that large and various school of collectivist thinkers in the nineteenth century who are lumped together as Socialists.

What is Socialism? There are a hundred definitions of Socialism and a thousand sects of Socialists. Essentially Socialism is no more and no less than a criticism of the idea of property in the light of the public good. We may review the history of that idea through the ages very briefly. That and the idea of internationalism are the two cardinal ideas upon which most of our political life is turning.

Photo: Linde & Co.

CARL MARX

The idea of property arises out of the combative instincts of the species. Long before men were men, the ancestral ape was a proprietor. Primitive property is what a beast will fight for. The dog and his bone, the tigress and her lair, the roaring stag and his herd, these are proprietorship blazing. No more nonsensical expression is conceivable in sociology than the term "primitive communism." The Old Man of the family tribe of early palæolithic times insisted upon his proprietorship in his wives and daughters, in his tools, in his visible universe. If any other man wandered into his visible universe he fought him, and if he could he slew him. The tribe grew in the course of ages, as Atkinson showed convincingly in his *Primal Law*, by the gradual toleration by the Old Man of the existence of the younger men, and of their proprietorship in the wives they captured from outside the tribe, and in the tools and ornaments they made and the game they slew. Human society grew by a compromise between this one's property and that. It was a compromise with instinct which was forced upon men by the necessity of driving some other tribe out of its visible universe. If the hills and forests and streams were not *your* land or *my* land, it was because they had to be *our* land. Each of us would have preferred to have it *my* land, but that would not work. In that case the other fellows would have destroyed us. Society, therefore, is from its beginning a *mitigation of ownership*. Ownership in the beast and in the primitive savage was far more intense a thing than it is in the civilized world to-day. It is rooted more strongly in our instincts than in our reason.

In the natural savage and in the untutored man to-day there is no limitation to the sphere of ownership. Whatever you can fight for, you can own; women-folk, spared captive, captured beast, forest glade, stone-pit or what not. As the community grew, a sort of law came to restrain internecine fighting, men developed rough-and-ready methods of settling proprietorship. Men could own what they were the first to make or capture or claim. It seemed natural that a debtor who could not pay should become the property of his creditor. Equally natural was it that after claiming a patch of land a man should exact payments from anyone who wanted to use it. It was only slowly, as the possibilities of organized life dawned on men, that this unlimited property in anything whatever began to be recognized as a nuisance. Men found themselves born into a universe all owned and claimed, nay! they found themselves born owned and claimed. The social struggles of the earlier civilization are difficult to trace now, but the history we have told of the Roman Republic shows a community waking up to the idea that debts may become a public inconvenience and should then be repudiated, and that the unlimited ownership of land is also an inconvenience. We find that later Babylonia severely limited the rights of property in slaves. Finally, we find in the teaching of that great revolutionist, Jesus of Nazareth, such an attack upon property as had never been before. Easier it was, he said, for a camel to go through the eye of a needle than for the owner of great possessions to enter the kingdom of heaven. A steady, continuous criticism of the permissible scope of property seems to have been going on in the world for the last twenty-five or thirty centuries. Nineteen hundred years after Jesus of Nazareth we find all the world that has come under the Christian teaching persuaded that there could be no property in human beings. And also the idea that "a man may do what he likes with his own" was very much shaken in relation to other sorts of property.

But this world of the closing eighteenth century was still only in the interrogative stage in this matter. It had got nothing clear enough, much less settled enough, to act upon. One of its primary impulses was to protect property against the greed and waste of kings and the exploitation of noble adventurers. It was largely to protect private property from taxation that the French Revolution began. But the equalitarian formulæ of the Revolution carried it into a criticism of the very property it had risen to protect. How can men be free and equal when numbers of them have no ground to stand upon and nothing to eat, and the owners will neither feed nor lodge them unless they toil? Excessively — the poor complained.

To which riddle the reply of one important political group was to set about "dividing up." They wanted to intensify and universalize property. Aiming at the same end by another route, there were the primitive socialists — or, to be more exact, communists — who wanted to "abolish" private property altogether. The state (a democratic state was of course understood) was to own all property.

It is paradoxical that different men seeking the same ends of liberty and happiness should propose on the one hand to make property as absolute as possible, and on the other to put an end to it altogether. But so it was. And the clue to this paradox is to be found in the fact that ownership is not one thing but a multitude of different things.

It was only as the nineteenth century developed that men began to realize that property was not one simple thing, but a great complex of ownerships of different values and consequences, that many things (such as one's body, the implements of an artist, clothing, toothbrushes) are very profoundly and incurably one's personal property, and that there is a very great range of things, railways, machinery of various sorts, homes, cultivated gardens, pleasure boats, for example, which need each to be considered very particularly to determine how far and under what limitations it may come under private ownership, and how far it falls into the public domain and may be administered and let out by the state in the collective interest. On the practical side these questions pass into politics, and the problem of making and sustaining efficient state administration. They open up issues in social psychology, and interact with the enquiries of educational science. The criticism of property is still a vast and passionate ferment rather than a science. On the one hand are the Individualists, who would protect and enlarge our present freedoms with what we possess, and on the other the Socialists who would in many directions pool our ownerships and restrain our proprietory acts. In practice one will find every gradation between the extreme individualist, who will scarcely tolerate a tax of any sort to support a government, and the communist who would deny any possessions at all. The ordinary socialist of to-day is what is called a collectivist; he would allow a considerable amount of private property but put such affairs as education, transport, mines, land-owning, most mass productions of staple articles, and the like, into the hands of a highly organized state. Nowadays there does seem to be a gradual convergence of reasonable men towards a moderate socialism scientifically studied and planned. It is realized more and more clearly that the untutored man does not co-operate easily and successfully in large undertakings, and that every step towards a more complex state and every function that the state takes over from private enterprise, necessitates a corresponding educational advance and the organization of a proper criticism and control. Both the press and the political methods of the contemporary state are far too crude for any large extension of collective activities.

But for a time the stresses between employer and employed and particularly between selfish employers and reluctant workers, led to a world-wide dissemination of the very harsh and elementary form of communism which is associated with the name of Marx. Marx based his theories on a belief that men's minds are limited by their economic necessities, and that there is a necessary conflict of interests in our present civilization between the prosperous and employing classes of people and the employed mass. With the

Photo: Jeffrey Manufacturing Company, Columbus, Ohio

SCIENCE IN THE COAL MINE
Portable Electric Loading Conveyor

advance in education necessitated by the mechanical revolution, this great employed majority will become more and more class-conscious and more and more solid in antagonism to the (class-conscious) ruling minority. In some way the class-conscious workers would seize power, he prophesied, and inaugurate a new social state. The antagonism, the insurrection, the possible revolution are understandable enough, but it does not follow that a new social state or anything but a socially destructive process will ensue. Put to the test in Russia, Marxism, as we shall note later, has proved singularly uncreative.

Marx sought to replace national antagonism by class antagonisms; Marxism has produced in succession a First, a Second and a Third Workers' International. But from the starting point of modern individualistic thought it is also possible to reach international ideas. From the days of that great English economist, Adam Smith, onward there has been an increasing realization that for world-wide prosperity free and unencumbered trade about the earth is needed. The individualist with his hostility to the state is hostile also to tariffs and boundaries and all the restraints upon free act and movement that national boundaries seem to justify. It is interesting to see two lines of thought, so diverse in spirit, so different in substance as this class-war socialism of the Marxists and the individualistic free-trading philosophy of the British business men of the Victorian age heading at last, in spite of these primary differences, towards the same intimations of a new world-wide treatment of human affairs outside the boundaries and limitations of any existing state. The logic of reality triumphs over the logic of theory. We begin to perceive that from widely divergent starting points individualist theory and socialist theory are part of a common search, a search for more spacious social and political ideas and interpretations, upon which men may contrive to work together, a search that began again in Europe and has intensified as men's confidence in the ideas of the Holy Roman Empire and in Christendom decayed, and as the age of discovery broadened their horizons from the world of the Mediterranean to the whole wide world.

To bring this description of the elaboration and development of social, economic and political ideas right down to the discussions of the present day, would be to introduce issues altogether too controversial for the scope and intentions of this book. But regarding these things, as we do here, from the vast perspectives of the student of world history, we are bound to recognize that this reconstruction of these directive ideas in the human mind is still an unfinished task —we cannot even estimate yet how unfinished the task may be. Certain common beliefs do seem to be emerging, and their influence is very perceptible upon the political events and public acts of to-day; but at present they are not clear enough nor convincing enough to compel men definitely and systematically towards their realization. Men's acts waver between tradition and the new, and on the whole they rather gravitate towards the traditional. Yet, compared with the thought of even a brief lifetime ago, there does seem to be an outline shaping itself of a new order in human affairs. It is a sketchy outline, vanishing into vagueness at this point and that,

Photo: Baker & Hurtzig

CONSTRUCTIONAL DETAIL OF THE FORTH BRIDGE

and fluctuating in detail and formulæ, yet it grows steadfastly clearer, and its main lines change less and less.

It is becoming plainer and plainer each year that in many respects and in an increasing range of affairs, mankind is becoming one community, and that it is more and more necessary that in such matters there should be a common world-wide control. For example, it is steadily truer that the whole planet is now one economic community, that the proper exploitation of its natural resources demands one comprehensive direction, and that the greater power and range that discovery has given human effort makes the present fragmentary and contentious administration of such affairs more and more wasteful and dangerous. Financial and monetary expedients also become world-wide interests to be dealt with successfully only on world-wide lines. Infectious diseases and the increase and migrations of population are also now plainly seen to be world-wide concerns. The greater power and range of human activities has also made war disproportionately destructive and disorganizing, and, even as a clumsy way of settling issues between government and government and people and people, ineffective. All these things clamour for controls and authorities of a greater range and greater comprehensiveness than any government that has hitherto existed.

But it does not follow that the solution of these problems lies in some super-government of all the world arising by conquest or by the coalescence of existing governments. By analogy with existing institutions men have thought of the Parliament of Mankind, of a World Congress, of a President or Emperor of the Earth. Our first natural reaction is towards some such conclusion, but the discussion and experiences of half a century of suggestions and attempts has on the whole discouraged belief in that first obvious idea. Along that line to world unity the resistances are too great. The drift of thought seems now to be in the direction of a number of special committees or organizations, with world-wide power delegated to them by existing governments in this group of matters or that, bodies concerned with the waste or development of natural wealth, with the equalization of labour conditions, with world peace, with currency, population and health, and so forth.

The world may discover that all its common interests are being managed as one concern, while it still fails to realize that a world government exists. But before even so much human unity is attained, before such international arrangements can be put above patriotic suspicions and jealousies, it is necessary that the common mind of the race should be possessed of that idea of human unity, and that the idea of mankind as one family should be a matter of universal instruction and understanding.

For a score of centuries or more the spirit of the great universal religions has been struggling to maintain and extend that idea of a universal human brotherhood, but to this day the spites, angers and distrusts of tribal, national and racial friction obstruct, and successfully obstruct, the broader views and more generous impulses which would make every man the servant of all mankind. The idea of human brotherhood struggles now to possess the human soul, just as the idea of Christendom struggled to possess the soul of Europe in the confusion and disorder of the sixth and seventh centuries of the Christian era. The dissemination and triumph of such ideas must be the work of a multitude of devoted and undistinguished missionaries, and no contemporary writer can presume to guess how far such work has gone or what harvest it may be preparing.

Social and economic questions seem to be inseparably mingled with international ones. The solution in each case lies in an appeal to that same spirit of service which can enter and inspire the human heart. The distrust, intractability and egotism of nations reflects and is reflected by the distrust, intractability and egotism of the individual owner and worker in the face of the common good. Exaggerations of possessiveness in the individual are parallel and of a piece with the clutching greed of nations and emperors. They are products of the same instinctive tendencies, and the same ignorances and traditions. Internationalism is the socialism of nations. No one who has wrestled with these problems can feel that there yet exists a sufficient depth and strength of psychological science and a sufficiently planned-out educational method and organization for any real and final solution of these riddles of human intercourse and co-operation. We are as incapable of planning a really effective peace organization of the world to-day as were men in 1820 to plan an electric railway system, but for all we know the thing is equally practicable and may be as nearly at hand.

No man can go beyond his own knowledge, no thought can reach beyond contemporary thought, and it is impossible for us to guess or foretell how many generations of humanity may have to live in war and waste and insecurity and misery before the dawn of the great peace to which all history seems to be pointing, peace in the heart and peace in the world, ends our night of wasteful and aimless living. Our proposed solutions are still vague and crude. Passion and suspicion surround them. A great task of intellectual reconstruction is going on, it is still incomplete, and our conceptions grow clearer and more exact — slowly, rapidly, it is hard to tell which. But as they grow clearer they will gather power over the minds and imaginations of men. Their present lack of grip is due to their lack of assurance and exact rightness. They are misunderstood because they are variously and confusingly presented. But with precision and certainty the new vision of the world will gain compelling power. It may presently gain power very rapidly. And a great work of educational reconstruction will follow logically and necessarily upon that clearer understanding.

LX

The Expansion of the United States

THE region of the world that displayed the most immediate and striking results from the new inventions in transport was North America. Politically the United States embodied, and its constitution crystallized, the liberal ideas of the middle eighteenth century. It dispensed with state-church or crown, it would have no titles, it protected property very jealously as a method of freedom, and — the exact practice varied at first in the different states — it gave nearly every adult male citizen a vote. Its method of voting was barbarically crude, and as a consequence its political life fell very soon under the control of highly organized party machines, but that did not prevent the newly emancipated population developing an energy, enterprise and public spirit far beyond that of any other contemporary population.

Then came that acceleration of locomotion to which we have already called attention. It is a curious thing that America, which owes most to this acceleration in locomotion, has felt it least. The United States have taken the railway, the river steamboat, the telegraph and so forth as though they were a natural part of their growth. They were not. These things happened to come along just in time to save American unity. The United States of to-day were made first by the river steamboat, and then by the railway. Without these things, the present United States, this vast continental nation, would have been altogether impossible. The westward flow of population would have been far more sluggish. It might never have crossed the great central plains. It took nearly two hundred years for effective settlement to reach from the coast to Missouri, much less than halfway across the continent. The first state established beyond the river was the steamboat state of Missouri in 1821. But the rest of the distance to the Pacific was done in a few decades.

If we had the resources of the cinema it would be interesting to show a map of North America year by year from 1600 onward, with little dots to represent hundreds of people, each dot a hundred, and stars to represent cities of a hundred thousand people.

For two hundred years the reader would see that stippling creeping slowly along the coastal districts and navigable waters, spreading still more gradually into Indiana, Kentucky and so forth. Then somewhere about 1810 would come a change. Things would get more lively along the river courses. The dots would be multiplying and spreading. That would be the steamboat. The pioneer dots would be spreading soon over Kansas and Nebraska from a number of jumping-off places along the great rivers.

Then from about 1830 onward would come the black lines of the railways, and after that the little black dots would not simply creep but run. They would appear now so rapidly, it would be almost as though they were being put on by some sort of spraying machine. And suddenly here and then there would appear the first stars to indicate the first great cities of a hundred thousand people. First one or two and then a multitude of cities — each like a knot in the growing net of the railways.

The growth of the United States is a process that has no precedent in the world's history; it is a new kind of occurrence. Such a community could not have come into existence before, and if it had, without railways it would certainly have dropped to pieces long before now. Without railways or telegraph it would be far easier to administer California from Pekin than from Washington. But this great population of the United States of America has not only grown outrageously; it has kept uniform. Nay, it has become more uniform. The man of San Francisco is more like the man of New York to-day than the man of Virginia was like the man of New England a century ago. And the process of assimilation goes on unimpeded. The United States is being woven by railway, by telegraph, more and more into one vast unity, speaking, thinking and acting harmoniously with itself. Soon aviation will be helping in the work.

This great community of the United States is an altogether new thing in history. There have been great empires before with populations exceeding 100 millions, but these were associations of divergent peoples; there has never been one single people on this scale before. We want a new term for this new thing. We call the United States a country just as we call France or Holland a country. But the two things are as different as an automobile and a one-horse shay. They are the creations of different periods and different conditions; they are going to work at a different pace and in an entirely different way. The United States in scale and possibility is halfway between a European state and a United States of all the world.

But on the way to this present greatness and security the American people passed through one phase of dire conflict. The river steamboats, the railways, the telegraph, and their associate facilities, did not come soon enough to avert a deepening conflict of interests and ideas between the southern and northern states of the Union. The former were slave-holding states; the latter, states in which all men were free. The railways and steamboats at first did but bring into sharper conflict an already established difference between the two sections of the United States. The increasing unification due to the new means of transport made the question whether the southern spirit or the northern should prevail an ever more urgent one. There was little possibility of compromise. The northern spirit was free and individualistic; the southern made for great estates and a conscious gentility ruling over a dusky subject multitude.

Every new territory that was organized into a state as the tide of population swept westward, every new incorporation into the fast growing American system, became a field of conflict between the two ideas, whether it should become a state of free citizens, or whether the estate and slavery system should prevail. From 1833 an American anti-slavery society was not merely resisting the extension of the institution but agitating the whole country for its complete abolition. The issue flamed up into open conflict over the admission of Texas to the Union. Texas had originally been a part of the republic of Mexico, but it was largely colonized by Americans from the slave-holding states, and it seceded from Mexico, established its independence in 1835, and was annexed to the United States in 1844. Under the Mexican law slavery had been forbidden in Texas, but now the South claimed Texas for slavery and got it.

Meanwhile the development of ocean navigation was bringing a growing swarm of immigrants from Europe to swell the spreading population of the northern states, and the raising of Iowa, Wisconsin, Minnesota and Oregon, all northern farm lands, to state level, gave the anti-slavery North the possibility of predominance both in the Senate and the House of Representatives. The cotton-growing South, irritated by the growing threat of the Abolitionist movement, and fearing this predominance in Congress, began to talk of secession from the Union. Southerners began to dream of annexations to the south of them in Mexico and the West Indies, and of a great slave state, detached from the North and reaching to Panama.

The return of Abraham Lincoln as an anti-extension President in 1860 decided the South to split the Union. South Carolina passed an "ordinance of secession," and prepared for war. Mississippi, Florida, Alabama, Georgia, Louisiana and Texas joined her, and a convention met at Montgomery in Alabama, elected Jefferson Davis president of the "Confederated States" of America, and adopted a constitution specifically upholding "the institution of negro slavery."

ONE OF THE FIRST AMERICAN RIVER STEAMERS

Abraham Lincoln was, it chanced, a man entirely typical of the new people that had grown up after the War of Independence. His early years had been spent as a drifting particle in the general westward flow of the population. He was born in Kentucky (1809), was taken to Indiana as a boy and later on to Illinois. Life was rough in the backwoods of Indiana in those days; the house was a mere log cabin in the wilderness, and his schooling was poor and casual. But his mother taught him to read early, and he became a voracious reader. At seventeen he was a big athletic youth, a great wrestler and runner. He worked for a time as clerk in a store, went into business as a storekeeper with a drunken partner, and contracted debts that he did not fully pay off for fifteen years. In 1834, when he was still only five and twenty, he was elected member of the House of Representatives for the State of Illinois. In Illinois particularly the question of slavery flamed because the great leader of the party for the extension of slavery in the national Congress was Senator Douglas of Illinois. Douglas was a man of great ability and prestige, and for some years Lincoln fought against him by speech and pamphlet, rising steadily to the position of his most formidable and finally victorious antagonist. Their culminating struggle was the presidential campaign of 1860, and on the fourth of March, 1861, Lincoln was inaugurated President, with the southern states already in active secession from the rule of the federal government at Washington, and committing acts of war.

This civil war in America was fought by improvised armies that grew steadily from a few score thousands to hundreds of thousands — until at last the Federal forces exceeded a million men; it was fought over a vast area between New Mexico and the eastern sea, Washington and Richmond were the chief objectives. It is beyond our scope here to tell of the mounting energy of that epic struggle that rolled to and fro across the hills and woods of Tennessee and Virginia and down the Mississippi. There was a terrible waste and killing of men. Thrust was followed by counter thrust; hope gave way to despondency, and returned and was again disappointed. Sometimes Washington seemed within the Confederate grasp; again the Federal armies were driving towards Richmond. The Confederates, outnumbered and far poorer in resources, fought under a general of supreme ability, General Lee. The generalship of the Union was far inferior. Generals were dismissed, new generals appointed; until at last, under Sherman and Grant, came victory

ABRAHAM LINCOLN

over the ragged and depleted South. In October, 1864, a Federal army under Sherman broke through the Confederate left and marched down from Tennessee through Georgia to the coast, right across the Confederate country, and then turned up through the

Carolinas, coming in upon the rear of the Confederate armies. Meanwhile Grant held Lee before Richmond until Sherman closed on him. On April 9th, 1865, Lee and his army surrendered at Appomattox Court House, and within a month all the remaining secessionist armies had laid down their arms and the Confederacy was at an end.

This four years' struggle had meant an enormous physical and moral strain for the people of the United States. The principle of state autonomy was very dear to many minds, and the North seemed in effect to be forcing abolition upon the South. In the border states brothers and cousins, even fathers and sons, would take opposite sides and find themselves in antagonistic armies. The North felt its cause a righteous one, but for great numbers of people it was not a full-bodied and unchallenged righteousness. But for Lincoln there was no doubt. He was a clear-minded man in the midst of much confusion. He stood for union; he stood for the wide peace of America. He was opposed to slavery, but slavery he held to be a secondary issue; his primary purpose was that the United States should not be torn into two contrasted and jarring fragments.

When in the opening stages of the war Congress and the Federal generals embarked upon a precipitate emancipation, Lincoln opposed and mitigated their enthusiasm. He was for emancipation by stages and with compensation. It was only in January, 1865, that the situation had ripened to a point when Congress could propose to abolish slavery for ever by a constitutional amendment, and the war was already over before this amendment was ratified by the states.

As the war dragged on through 1862 and 1863, the first passions and enthusiasms waned, and America learnt all the phases of war weariness and war disgust. The President found himself with defeatists, traitors, dismissed generals, tortuous party politicians, and a doubting and fatigued people behind him and uninspired generals and depressed troops before him; his chief consolation must have been that Jefferson Davis at Richmond could be in little better case. The English government misbehaved, and permitted the Confederate agents in England to launch and man three swift privateer ships — the *Alabama* is the best remembered of them — which chased United States shipping from the seas. The French army in Mexico was trampling the Monroe Doctrine in the dirt. Came subtle proposals from Richmond to drop the war, leave the issues of the war for subsequent discussion, and turn, Federal and Confederate in alliance, upon the French in Mexico. But Lincoln would not listen to such proposals unless the supremacy of the Union was maintained. The Americans might do such things as one people but not as two.

He held the United States together through long weary months of reverses and ineffective effort, through black phases of division and failing courage; and there is no record that he ever faltered from his purpose. There were times when there was nothing to be done, when he sat in the White House silent and motionless, a grim monument of resolve; times when he relaxed his mind by jesting and broad anecdotes.

He saw the Union triumphant. He entered Richmond the day after its surrender, and heard of Lee's capitulation. He returned to Washington, and on April 11th made his last public address. His theme was reconciliation and the reconstruction of loyal government in the defeated states. On the evening of April 14th he went to Ford's theatre in Washington, and as he sat looking at the stage, he was shot in the back of the head and killed by an actor named Booth who had some sort of grievance against him, and who had crept into the box unobserved. But Lincoln's work was done; the Union was saved.

At the beginning of the war there was no railway to the Pacific coast; after it the railways spread like a swiftly growing plant until now they have clutched and held and woven all the vast territory of the United States into one indissoluble mental and material unity — the greatest real community — until the common folk of China have learnt to read — in the world.

LXI

The Rise of Germany to Predominance in Europe

We have told how after the convulsion of the French Revolution and the Napoleonic adventure, Europe settled down again for a time to an insecure peace and a sort of modernized revival of the political conditions of fifty years before. Until the middle of the century the new facilities in the handling of steel and the railway and steamship produced no marked political consequences. But the social tension due to the development of urban industrialism grew. France remained a conspicuously uneasy country. The revolution of 1830 was followed by another in 1848. Then Napoleon III, a nephew of Napoleon Bonaparte, became first President, and then (in 1852) Emperor.

He set about rebuilding Paris, and changed it from a picturesque seventeenth century insanitary city into the spacious Latinized city of marble it is to-day. He set about rebuilding France, and made it into a brilliant-looking modernized imperialism. He displayed a disposition to revive that competitiveness of the Great Powers which had kept Europe busy with futile wars during the seventeenth and eighteenth centuries. The Tsar Nicholas I of Russia (1825–1856) was also becoming aggressive and pressing southward upon the Turkish Empire with his eyes on Constantinople.

After the turn of the century Europe broke out into a fresh cycle of wars. They were chiefly "balance-of-power" and ascendancy wars. England, France and Sardinia assailed Russia in the Crimean war in defence of Turkey; Prussia (with Italy as an ally) and Austria fought for the leadership of Germany, France liberated North Italy from Austria at the price of Savoy, and Italy gradually unified itself into one kingdom. Then Napoleon III was so ill advised as to attempt adventures in Mexico, during the American Civil War; he set up an Emperor Maximilian there and abandoned him hastily to his fate — he was shot by the Mexicans — when the victorious Federal Government showed its teeth.

In 1870 came a long-pending struggle for predominance in Europe between France and Prussia. Prussia had long foreseen and

prepared for this struggle, and France was rotten with financial corruption. Her defeat was swift and dramatic. The Germans invaded France in August, one great French army under the Emperor capitulated at Sedan in September, another surrendered in October at Metz, and in January 1871, Paris, after a siege and bombardment, fell into German hands. Peace was signed at Frankfort surrendering the provinces of Alsace and Lorraine to the Germans.

Germany, excluding Austria, was unified as an empire, and the King of Prussia was added to the galaxy of European Cæsars, as the German Emperor.

For the next forty-three years Germany was the leading power upon the European continent. There was a Russo-Turkish war in 1877–8, but thereafter, except for certain readjustments in the Balkans, European frontiers remained uneasily stable for thirty years.

LXII

The New Overseas Empires of Steamship and Railway

THE end of the eighteenth century was a period of disrupting empires and disillusioned expansionists. The long and tedious journey between Britain and Spain and their colonies in America prevented any really free coming and going between the home land and the daughter lands, and so the colonies separated into new and distinct communities, with distinctive ideas and interests and even modes of speech. As they grew they strained more and more at the feeble and uncertain link of shipping that had joined them. Weak trading-posts in the wilderness, like those of France in Canada, or trading establishments in great alien communities, like those of Britain in India, might well cling for bare existence to the nation which gave them support and a reason for their existence. That much and no more seemed to many thinkers in the early part of the nineteenth century to be the limit set to overseas rule. In 1820 the sketchy great European "empires" outside of Europe that had figured so bravely in the maps of the middle eighteenth century, had shrunken to very small dimensions. Only the Russian sprawled as large as ever across Asia.

The British Empire in 1815 consisted of the thinly populated coastal river and lake regions of Canada, and a great hinterland of wilderness in which the only settlements as yet were the fur-trading stations of the Hudson Bay Company, about a third of the Indian peninsula, under the rule of the East India Company, the coast districts of the Cape of Good Hope inhabited by blacks and rebellious-spirited Dutch settlers; a few trading stations on the coast of West Africa, the rock of Gibraltar, the island of Malta, Jamaica, a few minor slave-labour possessions in the West Indies, British Guiana in South America, and, on the other side of the world, two dumps for convicts at Botany Bay in Australia and in Tasmania. Spain retained Cuba and a few settlements in the Philippine Islands.

Portugal had in Africa some vestiges of her ancient claims. Holland had various islands and possessions in the East Indies and Dutch Guiana, and Denmark an island or so in the West Indies. France had one or two West Indian islands and French Guiana. This seemed to be as much as the European powers needed, or were likely to acquire of the rest of the world. Only the East India Company showed any spirit of expansion.

While Europe was busy with the Napoleonic wars the East India Company, under a succession of Governors-General, was playing much the same rôle in India that had been played before by Turkoman and such-like invaders from the north. And after the peace of Vienna it went on, levying its revenues, making wars, sending ambassadors to Asiatic powers, a quasi-independent state, however, with a marked disposition to send wealth westward.

We cannot tell here in any detail how the British Company made its way to supremacy sometimes as the ally of this power, sometimes as that, and finally as the conqueror of all. Its power spread to Assam, Sind, Oudh. The map of India began to take on the outlines familiar to the English schoolboy of to-day, a patchwork of native states embraced and held together by the great provinces under direct British rule. . . .

In 1859, following upon a serious mutiny of the native troops in India, this empire of the East India Company was annexed to the British Crown. By an Act entitled *An Act for the Better Government of India,* the Governor-General became a Viceroy representing the Sovereign, and the place of the Company was taken by a Secretary of State for India responsible to the British Parliament. In 1877, Lord Beaconsfield, to complete the work, caused Queen Victoria to be proclaimed Empress of India.

Upon these extraordinary lines India and Britain are linked at the present time. India is still the empire of the Great Mogul, but the Great Mogul has been replaced by the "crowned republic" of Great Britain. India is an autocracy without an autocrat. Its rule combines the disadvantage of absolute monarchy with the impersonality and irresponsibility of democratic officialdom. The Indian with a complaint to make has no visible monarch to go to; his Emperor is a golden symbol; he must circulate pamphlets in England or inspire a question in the British House of Commons. The more occupied Parliament is with British affairs, the less attention India will receive, and the more she will be at the mercy of her small group of higher officials.

Photo: British South African Co.
RAILWAY BRIDGE OVER THE GORGE, VICTORIA FALLS, OF THE ZAMBESI, SOUTHERN RHODESIA

Apart from India, there was no great expansion of any European Empire until the railways and the steamships were in effective action. A considerable school of political thinkers in Britain was disposed to regard overseas possessions as a source of weakness to the kingdom. The Australian settlements developed slowly until in 1842 the discovery of valuable copper mines, and in 1851 of gold, gave them a new importance. Improvements in transport were also making Australian wool an increasingly marketable commodity in Europe.

Canada, too, was not remarkably progressive until 1849; it was troubled by dissensions between its French and British inhabitants, there were several serious revolts, and it was only in 1867 that a new constitution creating a Federal Dominion of Canada relieved its internal strains. It was the railway that altered the Canadian outlook. It enabled Canada, just as it enabled the United States, to expand westward, to market its corn and other produce in Europe, and in spite of its swift and extensive growth, to remain in language and sympathy and interests one community. The railway, the steamship and the telegraph cable were indeed changing all the conditions of colonial development.

Before 1840, English settlements had already begun in New Zealand, and a New Zealand Land Company had been formed to exploit the possibilities of the island. In 1840 New Zealand also was added to the colonial possessions of the British Crown.

Canada, as we have noted, was the first of the British possessions to respond richly to the new economic possibilities that the new methods of transport were opening. Presently the republics of South America, and particularly the Argentine Republic, began to feel in their cattle trade and coffee growing the increased nearness of the European market. Hitherto the chief commodities that had attracted the European powers into unsettled and barbaric regions had been gold or other metals, spices, ivory, or slaves. But in the latter quarter of the nineteenth century the increase of the European populations was obliging their governments to look abroad for staple foods; and the growth of scientific industrialism was creating a demand for new raw materials, fats and greases of every kind, rubber, and other hitherto disregarded substances. It was plain that Great Britain and Holland and Portugal were reaping a great and growing commercial advantage from their very considerable control of tropical and sub-tropical products. After 1871 Germany, and presently France and later Italy, began to look for unannexed raw-material areas, or for Oriental countries capable of profitable modernization.

So began a fresh scramble all over the world, except in the American region where the Monroe Doctrine now barred such adventures, for politically unprotected lands.

Close to Europe was the continent of Africa, full of vaguely known possibilities. In 1850 it was a continent of black mystery; only Egypt and the coast were known. Here we have no space to tell the amazing story of the explorers and adventurers who first pierced the African darkness, and of the political agents, administrators, traders, settlers and scientific men who followed in their track. Wonderful races of men like the pygmies, strange beasts like the okapi, marvellous fruits and flowers and insects, terrible diseases, astounding scenery of forest and mountain, enormous inland seas and gigantic rivers and cascades were revealed; a whole new world. Even remains (at Zimbabwe) of some unrecorded and vanished civilization, the southward enterprise of an early people, were discovered. Into this new world came the Europeans, and found the rifle already there in the hands of the Arab slave-traders, and negro life in disorder.

By 1900, in half a century, all Africa was mapped, explored, estimated and divided between the European powers. Little heed was given to the welfare of the natives in this scramble. The Arab slaver was indeed curbed rather than expelled, but the greed for rubber, which was a wild product collected under compulsion by the natives in the Belgian Congo, a greed exacerbated by the clash of inexperienced European administrators with the native population, led to horrible atrocities. No European power has perfectly clean hands in this matter.

We cannot tell here in any detail how Great Britain got possession of Egypt in 1883 and remained there in spite of the fact that Egypt was technically a part of the Turkish Empire, nor how nearly this scramble led to war between France and Great Britain in 1898, when a certain Colonel Marchand, crossing Central Africa from the west coast, tried at Fashoda to seize the Upper Nile.

Nor can we tell how the British Government first let the Boers, or Dutch settlers, of the Orange River district and the Transvaal set up independent republics in the inland parts of South Africa, and then repented and annexed the Transvaal Republic in 1877; nor how the Transvaal Boers fought for freedom and won it after the battle of Majuba Hill (1881). Majuba Hill was made to rankle in the memory of the English people by a persistent press campaign. A war with both republics broke out in 1899, a three years' war enormously costly to the British people, which ended at last in the surrender of the two republics.

Their period of subjugation was a brief one. In 1907, after the downfall of the imperialist government which had conquered them, the Liberals took the South African problem in hand, and these former republics became free and fairly willing associates with Cape Colony and Natal in a Confederation of all the states of South Africa as one self-governing republic under the British Crown.

In a quarter of a century the partition of Africa was completed. There remained unannexed three comparatively small countries: Liberia, a settlement of liberated negro slaves on the west coast; Morocco, under a Moslem Sultan; and Abyssinia, a barbaric country, with an ancient and peculiar form of Christianity, which had successfully maintained its independence against Italy at the battle of Adowa in 1896.

LXIII

European Aggression in Asia, and the Rise of Japan

IT is difficult to believe that any large number of people really accepted this headlong painting of the map of Africa in European colours as a permanent new settlement of the world's affairs, but it is the duty of the historian to record that it was so accepted. There was but a shallow historical background to the European mind in the nineteenth century, and no habit of penetrating criticism. The quite temporary advantages that the mechanical revolution in the west had given the Europeans over the rest of the old world were regarded by people, blankly ignorant of such events as the great Mongol conquests, as evidences of a permanent and assured European leadership of mankind. They had no sense of the transferability of science and its fruits. They did not realize that Chinamen and Indians could carry on the work of research as ably as Frenchmen or Englishmen. They believed that there was some innate intellectual drive in the west, and some innate indolence and conservatism in the east, that assured the Europeans a world predominance for ever.

The consequence of this infatuation was that the various European foreign offices set themselves not merely to scramble with the British for the savage and undeveloped regions of the world's surface, but also to carve up the populous and civilized countries of Asia as though these people also were no more than raw material for exploitation. The inwardly precarious but outwardly splendid imperialism of the British ruling class in India, and the extensive and profitable possessions of the Dutch in the East Indies, filled the rival Great Powers with dreams of similar glories in Persia, in the disintegrating Ottoman Empire, and in Further India, China and Japan.

In 1898 Germany seized Kiau Chau in China. Britain responded by seizing Wei-hai-wei, and the next year the Russians took possession of Port Arthur. A flame of hatred for the Europeans swept through China. There were massacres of Europeans and Christian converts, and in 1900 an attack upon and siege of the European legations in Pekin. A combined force of Europeans made a punitive expedition to Pekin, rescued the legations, and stole an enormous amount of valuable property. The Russians then seized Manchuria, and in 1904 the British invaded Tibet. . . .

But now a new Power appeared in the struggle of the Great Powers, Japan. Hitherto Japan has played but a small part in this history; her secluded civilization has not contributed very largely to the general shaping of human destinies; she has received much, but she has given little. The Japanese proper are of the Mongolian race. Their civilization, their writing and their literary and artistic traditions are derived from the Chinese. Their history is an interesting and romantic one; they developed a feudal system and a system of chivalry in the earlier centuries of the Christian era; their attacks upon Korea and China are an Eastern equivalent of the English wars in France. Japan was first brought into contact with Europe in the sixteenth century; in 1542 some Portuguese reached it in a Chinese junk, and in 1549 a Jesuit missionary, Francis Xavier, began his teaching there. For a time Japan welcomed European intercourse, and the Christian missionaries made a great number of converts. A certain William Adams became the most trusted European adviser of the Japanese, and showed them how to build big ships. There were voyages in Japanese-built ships to India and Peru. Then arose complicated quarrels between the Spanish Dominicans, the Portuguese Jesuits, and the English and Dutch Protestants, each warning the Japanese against the political designs of the others. The Jesuits, in a phase of ascendancy, persecuted and insulted the Buddhists with great acrimony. In the end the Japanese came to the conclusion that the Europeans were an intolerable nuisance, and that Catholic Christianity in particular was a mere cloak for the political dreams of the Pope and the Spanish monarchy — already in possession of the Philippine Islands; there was a great persecution of the Christians, and in 1638 Japan was absolutely closed to Europeans, and remained closed for over 200 years. During those two centuries the Japanese were as completely cut off from the rest of the world as though they lived upon another planet. It was forbidden to build any ship larger than a mere coasting boat. No Japanese could go abroad, and no European enter the country.

JAPANESE SOLDIER OF THE EIGHTEENTH CENTURY
(In the Victoria and Albert Museum)

For two centuries Japan remained outside the main current of history. She lived on in a state of picturesque feudalism in which about five per cent. of the population, the *samurai*, or fighting men, and the nobles and their families, tyrannized without restraint over the rest of the population. Meanwhile the great world outside went on to wider visions and new powers. Strange shipping became more frequent, passing the Japanese headlands; sometimes ships were wrecked and sailors brought ashore. Through the Dutch settlement in the island of Deshima, their one link with the outer universe, came warnings that Japan was not keeping pace with the power of the Western world. In 1837 a ship sailed into Yedo Bay flying a strange flag of stripes and stars, and carrying some Japanese sailors she had picked up far adrift in the Pacific. She was driven off by cannon shot. This flag presently reappeared on other ships. One in 1849 came to demand the liberation of eighteen shipwrecked American sailors. Then in 1853 came four American warships under Commodore Perry, and refused to be driven away. He lay at anchor in forbidden waters, and sent messages to the two rulers who at that time shared the control of Japan. In 1854 he returned with ten ships, amazing ships propelled by steam, and equipped with big guns, and he made proposals for trade and intercourse that the Japanese had no power to resist. He landed with a guard of 500 men to sign the treaty. Incredulous crowds watched this visitation from the outer world, marching through the streets.

Russia, Holland and Britain followed in the wake of America. A great nobleman whose estates commanded the Straits of Shimonoseki saw fit to fire on foreign vessels, and a bombardment by a fleet of British, French, Dutch and American warships destroyed his batteries and scattered his swordsmen. Finally an allied squadron (1865), at anchor off Kioto, imposed a ratification of the treaties which opened Japan to the world.

The humiliation of the Japanese by these events was intense. With astonishing energy and intelligence they set themselves to bring their culture and organization to the level of the European Powers. Never in all the history of mankind did a nation make such a stride as Japan then did. In 1866 she was a medieval people, a fantastic caricature of the extremest romantic feudalism; in 1899 hers was a completely Westernized people, on a level with the most advanced European Powers. She completely dispelled the persuasion that Asia was in some irrevocable way hopelessly behind Europe. She made all European progress seem sluggish by comparison.

We cannot tell here in any detail of Japan's war with China in 1894–95. It demonstrated the extent of her Westernization. She had an efficient Westernized army and a small but sound fleet. But the significance of her renascence, though it was appreciated by Britain and the United States, who were already treating her as if she were a European state, was not understood by the other Great Powers engaged in the pursuit of new Indias in Asia. Russia was pushing down through Manchuria to Korea. France was already established far to the south in Tonkin and Annam, Germany was prowling hungrily on the look-out for some settlement. The three Powers combined to prevent Japan reaping any fruits from the Chinese war. She was exhausted by the struggle, and they threatened her with war.

Japan submitted for a time and gathered her forces. Within ten years she was ready for a struggle with Russia, which marks an epoch in the history of Asia, the close of the period of European

A STREET IN TOKIO

arrogance. The Russian people were, of course, innocent and ignorant of this trouble that was being made for them halfway round the world, and the wiser Russian statesmen were against these foolish thrusts; but a gang of financial adventurers, including the Grand Dukes, his cousins, surrounded the Tsar. They had gambled deeply in the prospective looting of Manchuria and China, and they would suffer no withdrawal. So there began a transportation of great armies of Japanese soldiers across the sea to Port Arthur and Korea, and the sending of endless trainloads of Russian peasants along the Siberian railway to die in those distant battlefields.

The Russians, badly led and dishonestly provided, were beaten on sea and land alike. The Russian Baltic Fleet sailed round Africa to be utterly destroyed in the Straits of Tshushima. A revolutionary movement among the common people of Russia, infuriated by this remote and reasonless slaughter, obliged the Tsar to end the war (1905); he returned the southern half of Saghalien, which had been seized by Russia in 1875, evacuated Manchuria, resigned Korea to Japan. The European invasion of Asia was coming to an end and the retraction of Europe's tentacles was beginning.

LXIV

The British Empire in 1914

WE may note here briefly the varied nature of the constituents of the British Empire in 1914 which the steamship and railway had brought together. It was and is a quite unique political combination; nothing of the sort has ever existed before.

First and central to the whole system was the "crowned republic" of the United British Kingdom, including (against the will of a considerable part of the Irish people) Ireland. The majority of the British Parliament, made up of the three united parliaments of England and Wales, Scotland and Ireland, determines the headship, the quality and policy of the ministry, and determines it largely on considerations arising out of British domestic politics. It is this ministry which is the effective supreme government, with powers of peace and war, over all the rest of the empire.

Next in order of political importance to the British States were the "crowned republics" of Australia, Canada, Newfoundland (the oldest British possession, 1583), New Zealand and South Africa, all practically independent and self-governing states in alliance with Great Britain, but each with a representative of the Crown appointed by the Government in office;

Next the Indian Empire, an extension of the Empire of the Great Mogul, with its dependent and "protected" states reaching now from Beluchistan to Burma, and including Aden, in all of which empire the British Crown *and* the India Office (under Parliamentary control) played the rôle of the original Turkoman dynasty;

Then the ambiguous possession of Egypt, still nominally a part of the Turkish Empire and still retaining its own monarch, the Khedive, but under almost despotic British official rule;

Then the still more ambiguous "Anglo-Egyptian" Sudan

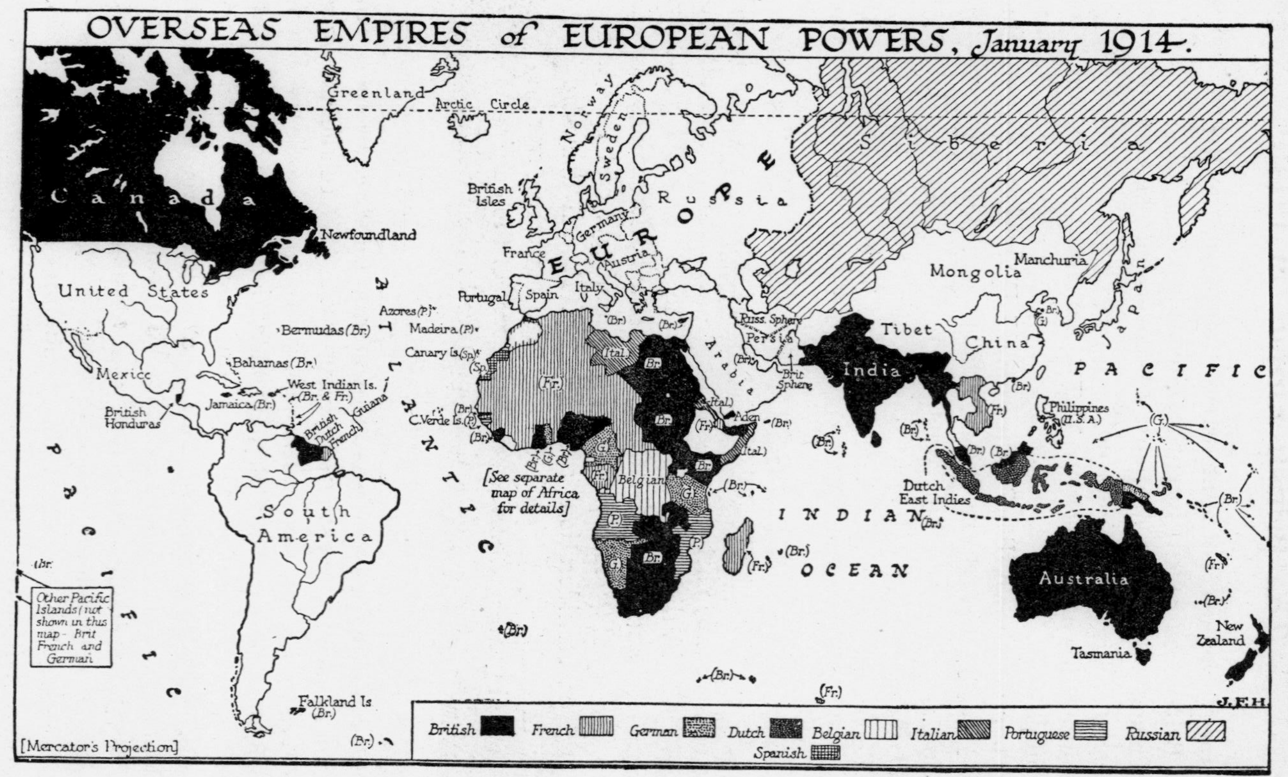
OVERSEAS EMPIRES of EUROPEAN POWERS, January 1914.
Canada
Greenland
Arctic Circle
Newfoundland
United States
Mexico
British Honduras
Bahamas (Br.)
Bermudas (Br.)
West Indian Is. (Br. & Fr.)
Jamaica (Br.)
Guiana
British
Dutch
French
South America
Falkland Is (Br.)
(Br.)
Other Pacific Islands (not shown in this map - Brit French and German
[Mercator's Projection]
PACIFIC
ATLANTIC
Azores (P.)
Madeira (P.)
Canary Is. (Sp.)
(Sp.)
C. Verde Is. (P.)
(Br.)
British Isles
Norway
Sweden
Germany
Austria
France
Portugal
Spain
Italy
EUROPE
Russia
Siberia
Mongolia
Manchuria
Japan
Tibet
China
India
Russ. Sphere
Persia
Brit Sphere
Arabia
Aden
(Fr.)
(Ital.)
(G.)
Ital.
Belgian
[See separate map of Africa for details]
INDIAN OCEAN
Philippines (U.S.A.)
Dutch East Indies
Australia
Tasmania
New Zealand
J.F.H.
British
French
German
Dutch
Belgian
Italian
Portuguese
Russian
Spanish

province, occupied and administered jointly by the British and by the (British controlled) Egyptian Government;

Then a number of partially self-governing communities, some British in origin and some not, with elected legislatures and an appointed executive, such as Malta, Jamaica, the Bahamas and Bermuda;

Then the Crown colonies, in which the rule of the British Home Government (through the Colonial Office) verged on autocracy, as in Ceylon, Trinidad and Fiji (where there was an appointed council), and Gibraltar and St. Helena (where there was a governor);

Then great areas of (chiefly) tropical lands, raw-product areas, with politically weak and under-civilized native communities which

Photo: C. Sinclair

GIBRALTAR

were nominally protectorates, and administered either by a High Commissioner set over native chiefs (as in Basutoland) or over a chartered company (as in Rhodesia). In some cases the Foreign Office, in some cases the Colonial Office, and in some cases the India Office, has been concerned in acquiring the possessions that fell into this last and least definite class of all, but for the most part the Colonial Office was now responsible for them.

It will be manifest, therefore, that no single office and no single brain had ever comprehended the British Empire as a whole. It was a mixture of growths and accumulations entirely different from anything that has ever been called an empire before. It guaranteed a wide peace and security; that is why it was endured and sustained by many men of the "subject" races — in spite of official tyrannies

Photo: Underwood & Underwood

STREET IN HONG KONG

and insufficiencies, and of much negligence on the part of the "home" public. Like the Athenian Empire, it was an overseas empire; its ways were sea ways, and its common link was the British Navy. Like all empires, its cohesion was dependent physically upon a method of communication; the development of seamanship, ship-building and steamships between the sixteenth and nineteenth centuries had made it a possible and convenient Pax — the "Pax Britannica," and fresh developments of air or swift land transport might at any time make it inconvenient.

LXV

The Age of Armament in Europe, and the Great War of 1914–18

THE progress in material science that created this vast steamboat-and-railway republic of America and spread this precarious British steamship empire over the world, produced quite other effects upon the congested nations upon the continent of Europe. They found themselves confined within boundaries fixed during the horse-and-high-road period of human life, and their expansion overseas had been very largely anticipated by Great Britain. Only Russia had any freedom to expand eastward; and she drove a great railway across Siberia until she entangled herself in a conflict with Japan, and pushed south-eastwardly towards the borders of Persia and India to the annoyance of Britain. The rest of the European Powers were in a state of intensifying congestion. In order to realize the full possibilities of the new apparatus of human life they had to rearrange their affairs upon a broader basis, either by some sort of voluntary union or by a union imposed upon them by some predominant power. The tendency of modern thought was in the direction of the former alternative, but all the force of political tradition drove Europe towards the latter.

The downfall of the "empire" of Napoleon III, the establishment of the new German Empire, pointed men's hopes and fears towards the idea of a Europe consolidated under German auspices. For thirty-six years of uneasy peace the politics of Europe centred upon that possibility. France, the steadfast rival of Germany for European ascendancy since the division of the empire of Charlemagne, sought to correct her own weakness by a close alliance with Russia, and Germany linked herself closely with the Austrian Empire (it had ceased to be the Holy Roman Empire in the days of Napoleon I) and less successfully with the new kingdom of Italy.

At first Great Britain stood as usual half in and half out of continental affairs. But she was gradually forced into a close association with the Franco-Russian group by the aggressive development of a great German navy. The grandiose imagination of the Emperor

Photo: British Official

BRITISH TANK IN THE BATTLE OF THE MENIN ROAD

The crew come out for a breath of fresh air during a lull

William II (1888–1918) thrust Germany into premature overseas enterprise that ultimately brought not only Great Britain but Japan and the United States into the circle of her enemies.

All these nations armed. Year after year the proportion of national production devoted to the making of guns, equipment, battleships and the like, increased. Year after year the balance

Photo: Topical

THE RUINS OF YPRES (ONCE A DELIGHTFUL OLD FLEMISH TOWN)

To show the complete destructiveness of modern war

of things seemed trembling towards war, and then war would be averted. At last it came. Germany and Austria struck at France and Russia and Serbia; the German armies marching through Belgium, Britain immediately came into the war on the side of Belgium, bringing in Japan as her ally, and very soon Turkey followed on the German side. Italy entered the war against Austria in 1915, and Bulgaria joined the Central Powers in the October of that year. In 1916 Rumania, and in 1917 the United States and China were forced into war against Germany. It is not within the scope of this history to define the exact share of blame for this vast catastrophe. The more interesting question is not why the Great War was begun but why the Great War was not anticipated and prevented. It is a far graver thing for mankind that scores of millions of people were too "patriotic," stupid, or apathetic to prevent this disaster by a movement towards European unity upon frank and generous lines, than that a small number of people may have been active in bringing it about.

It is impossible within the space at our command here to trace the intricate details of the war. Within a few months it became apparent that the progress of modern technical science had changed the nature of warfare very profoundly. Physical science gives power, power over steel, over distance, over disease; whether that power is used well or ill depends upon the moral and political intelligence of the world. The governments of Europe, inspired by antiquated policies of hate and suspicion, found themselves with unexampled powers both of destruction and resistance in their hands. The war became a consuming fire round and about the world, causing losses both to victors and vanquished out of all proportion to the issues involved. The first phase of the war was a tremendous rush of the Germans upon Paris and an invasion of East Prussia by the Russians. Both attacks were held and turned. Then the power of the defensive developed; there was a rapid elaboration of trench warfare until for a time the opposing armies lay entrenched in long lines right across Europe, unable to make any advance without enormous losses. The armies were millions strong, and behind them entire populations were organized for the supply of food and munitions to the front. There was a cessation of nearly every sort of productive activity except such as contributed to military operations. All the able-bodied manhood of Europe was drawn into the armies or navies or into the improvised

Photo: Photopress

THE DEVASTATION OF MODERN WAR

Wire entanglements in the foreground

factories that served them. There was an enormous replacement of men by women in industry. Probably more than half the people in the belligerent countries of Europe changed their employment altogether during this stupendous struggle. They were socially uprooted and transplanted. Education and normal scientific work were restricted or diverted to immediate military ends, and the distribution of news was crippled and corrupted by military control and "propaganda" activities.

The phase of military deadlock passed slowly into one of aggression upon the combatant populations behind the fronts by the destruction of food supplies and by attacks through the air. And also there was a steady improvement in the size and range of the guns employed and of such ingenious devices as poison-gas shells and the small mobile forts known as tanks, to break down the resistance of troops in the trenches. The air offensive was the most revolutionary of all the new methods. It carried warfare from two dimensions into three. Hitherto in the history of mankind war had gone on only where the armies marched and met. Now it went on everywhere. First the Zeppelin and then the bombing aeroplane carried war over and past the front to an ever-increasing area of civilian activities beyond. The old distinction maintained in civilized warfare between the civilian and combatant population disappeared. Everyone who grew food, or who sewed a garment, everyone who felled a tree or repaired a house, every railway station and every warehouse was held to be fair game for destruction. The air offensive increased in range and terror with every month in the war. At last great areas of Europe were in a state of siege and subject to nightly raids. Such exposed cities as London and Paris passed sleepless night after sleepless night while the bombs burst, the anti-aircraft guns maintained an intolerable racket, and the fire engines and ambulances rattled headlong through the darkened and deserted streets. The effects upon the minds and health of old people and of young children were particularly distressing and destructive.

Pestilence, that old follower of warfare, did not arrive until the very end of the fighting in 1918. For four years medical science staved off any general epidemic; then came a great outbreak of influenza about the world which destroyed many millions of people. Famine also was staved off for some time. By the beginning of 1918 however most of Europe was in a state of mitigated and regulated famine. The production of food throughout the world had fallen very greatly through the calling off of peasant mankind to the fronts, and the distribution of such food as was produced was impeded by the havoc wrought by the submarine, by the rupture of customary routes through the closing of frontiers, and by the disorganization of the transport system of the world. The various governments took possession of the dwindling food supplies, and, with more or less success, *rationed* their populations. By the fourth year the whole world was suffering from shortages of clothing and housing and of most of the normal gear of life as well as of food. Business and economic life were profoundly disorganized. Everyone was worried, and most people were leading lives of unwonted discomfort.

The actual warfare ceased in November, 1918. After a supreme effort in the spring of 1918 that almost carried the Germans to Paris, the Central Powers collapsed. They had come to an end of their spirit and resources.

LXVI

The Revolution and Famine in Russia

BUT a good year and more before the collapse of the Central Powers the half oriental monarchy of Russia, which had professed to be the continuation of the Byzantine Empire, had collapsed. The Tsardom had been showing signs of profound rottenness for some years before the war; the court was under the sway of a fantastic religious impostor, Rasputin, and the public administration, civil and military, was in a state of extreme inefficiency and corruption. At the outset of the war there was a great flare of patriotic enthusiasm in Russia. A vast conscript army was called up, for which there was neither adequate military equipment nor a proper supply of competent officers, and this great host, ill supplied and badly handled, was hurled against the German and Austrian frontiers.

There can be no doubt that the early appearance of Russian armies in East Prussia in September, 1914, diverted the energies and attention of the Germans from their first victorious drive upon Paris. The sufferings and deaths of scores of thousands of ill-led Russian peasants saved France from complete overthrow in that momentous opening campaign, and made all western Europe the debtors of that great and tragic people. But the strain of the war upon this sprawling, ill-organized empire was too heavy for its strength. The Russian common soldiers were sent into battle without guns to support them, without even rifle ammunition; they were wasted by their officers and generals in a delirium of militarist enthusiasm. For a time they seemed to be suffering mutely as the beasts suffer; but there is a limit to the endurance even of the most ignorant. A profound disgust for Tsardom was creeping through these armies of betrayed and wasted men. From the close of 1915 onward Russia was a source of deepening anxiety to her Western Allies. Throughout 1916 she remained largely on the defensive, and there were rumours of a separate peace with Germany.

On December 29th, 1916, the monk Rasputin was murdered at a dinner party in Petrograd, and a belated attempt was made to put the Tsardom in order. By March things were moving rapidly; food riots in Petrograd developed into a revolutionary insurrection; there was an attempted suppression of the Duma, the representative body, there were attempted arrests of liberal leaders, the formation of a provisional government under Prince Lvoff, and an abdication (March 15th) by the Tsar. For a time it seemed that a moderate and controlled revolution might be possible — perhaps under a new Tsar. Then it became evident that the destruction of popular confidence in Russia had gone too far for any such adjustments. The Russian people were sick to death of the old order of things in Europe, of Tsars and wars and of Great Powers; it wanted relief, and that speedily, from unendurable miseries. The Allies had no understanding of Russian realities; their diplomatists were ignorant of Russian, genteel persons with their attention directed to the Russian Court rather than to Russia, they blundered steadily with the new situation. There was little goodwill among these diplomatists for republicanism, and a manifest disposition to embarrass the new government as much as possible. At the head of the Russian republican government was an eloquent and picturesque leader, Kerensky, who found himself assailed by the forces of a profounder revolutionary movement, the "social revolution," at home and cold-shouldered by the Allied governments abroad. His Allies would neither let him give the Russian peasants the land for which they craved nor peace beyond their frontiers. The French and the British press pestered their exhausted ally for a fresh offensive, but when presently the Germans made a strong attack by sea and land upon Riga, the British Admiralty quailed before the prospect of a Baltic expedition in relief. The new Russian Republic had to fight unsupported. In spite of their naval predominance and the bitter protests of the great English admiral, Lord Fisher (1841–1920), it is to be noted that the British and their Allies, except for some submarine attacks, left the Germans the complete mastery of the Baltic throughout the war.

The Russian masses, however, were resolute to end the war. At any cost. There had come into existence in Petrograd a body representing the workers and common soldiers, the Soviet, and this body clamoured for an international conference of socialists at Stockholm. Food riots were occurring in Berlin at this time, war weariness in Austria and Germany was profound, and there can be little doubt, in the light of subsequent events, that such a conference would have precipitated a reasonable peace on democratic lines in 1917 and a German revolution. Kerensky implored his Western allies to allow this conference to take place, but, fearful of a world-wide outbreak of socialism and republicanism, they refused, in spite of the favourable response of a small majority of the British Labour Party. Without either moral or physical help from the Allies, the unhappy "moderate" Russian Republic still fought on and made a last desperate offensive effort in July. It failed after some preliminary successes, and there came another great slaughtering of Russians.

The limit of Russian endurance was reached. Mutinies broke out in the Russian armies, and particularly upon the northern front, and on November 7th, 1917, Kerensky's government was overthrown and power was seized by the Soviets, dominated by the Bolshevik socialists under Lenin, and pledged to make peace regardless of the Western powers. On March 2nd, 1918, a separate peace between Russia and Germany was signed at Brest-Litovsk.

It speedily became evident that these Bolshevik socialists were men of a very different quality from the rhetorical constitutionalists and revolutionaries of the Kerensky phase. They were fanatical Marxist communists. They believed that their accession to power in Russia was only the opening of a world-wide social revolution, and they set about changing the social and economic order with the thoroughness of perfect faith and absolute inexperience. The western European and the American governments were themselves much too ill-informed and incapable to guide or help this extraordinary experiment, and the press set itself to discredit and the ruling classes to wreck these usurpers upon any terms and at any cost to themselves or to Russia. A propaganda of abominable and disgusting inventions went on unchecked in the press of the

By courtesy of Messrs. Hodder & Stoughton

A VIEW IN PETERSBURG UNDER BOLSHEVIK RULE

A wooden house has been demolished for firewood

world; the Bolshevik leaders were represented as incredible monsters glutted with blood and plunder and living lives of sensuality before which the realities of the Tsarist court during the Rasputin regime paled to a white purity. Expeditions were launched at the exhausted country, insurgents and raiders were encouraged, armed and subsidized, and no method of attack was too mean or too monstrous for the frightened enemies of the Bolshevik regime. In 1919, the Russian Bolsheviks, ruling a country already exhausted and disorganized by five years of intensive warfare, were fighting a British Expedition at Archangel, Japanese invaders in Eastern Siberia, Roumanians with French and Greek contingents in the south, the Russian Admiral Koltchak in Siberia and General Deniken, supported by the French fleet, in the Crimea. In July of that year an Esthonian army, under General Yudenitch, almost got to Petersburg. In 1920 the Poles, incited by the French, made a new attack on Russia; and a new reactionary raider, General Wrangel, took over the task of General Deniken in invading and devastating his own country. In March, 1921, the sailors at Cronstadt revolted. The Russian Government under its president, Lenin, survived all these various attacks. It showed an amazing tenacity, and the common people of Russia sustained it unswervingly under conditions of extreme hardship. By the end of 1921 both Britain and Italy had made a sort of recognition of the communist rule.

But if the Bolshevik Government was successful in its struggle against foreign intervention and internal revolt, it was far less happy in its attempts to set up a new social order based upon communist ideas in Russia. The Russian peasant is a small land-hungry proprietor, as far from communism in his thoughts and methods as a whale is from flying; the revolution gave him the land of the great landowners but could not make him grow food for anything but negotiable money, and the revolution, among other things, had practically destroyed the value of money. Agricultural production, already greatly disordered by the collapse of the railways through war-strain, shrank to a mere cultivation of food by the peasants for their own consumption. The towns starved. Hasty and ill-planned attempts to make over industrial production in accordance with communist ideas were equally unsuccessful. By 1920 Russia presented the unprecedented spectacle of a modern civilization in complete collapse. Railways were rusting and passing out of use, towns were falling into ruin, everywhere there was an immense mortality. Yet the country still fought with its enemies at its gates. In 1921 came a drought and a great famine among the peasant cultivators in the war-devastated south-east provinces. Millions of people starved.

But the question of the distresses and the possible recuperation of Russia brings us too close to current controversies to be discussed here.

LXVII

The Political and Social Reconstruction of the World

THE scheme and scale upon which this History is planned do not permit us to enter into the complicated and acrimonious disputes that centre about the treaties, and particularly of the treaty of Versailles, which concluded the Great War. We are beginning to realize that that conflict, terrible and enormous as it was, ended nothing, began nothing and settled nothing. It killed millions of people; it wasted and impoverished the world. It smashed Russia altogether. It was at best an acute and frightful reminder that we were living foolishly and confusedly without much plan or foresight in a dangerous and unsympathetic universe. The crudely organized egotisms and passions of national and imperial greed that carried mankind into that tragedy, emerged from it sufficiently unimpaired to make some other similar disaster highly probable so soon as the world has a little recovered from its war exhaustion and fatigue. Wars and revolutions make nothing; their utmost service to mankind is that, in a very rough and painful way, they destroy superannuated and obstructive things. The great war lifted the threat of German imperialism from Europe, and shattered the imperialism of Russia. It cleared away a number of monarchies. But a multitude of flags still waves in Europe, the frontiers still exasperate, great armies accumulate fresh stores of equipment.

The Peace Conference at Versailles was a gathering very ill adapted to do more than carry out the conflicts and defeats of the war to their logical conclusions. The Germans, Austrians, Turks and Bulgarians were permitted no share in its deliberations; they were only to accept the decisions it dictated to them. From the point of view of human welfare the choice of the place of meeting was particularly unfortunate. It was at Versailles in 1871 that, with every circumstance of triumphant vulgarity, the new German

Empire had been proclaimed. The suggestion of a melodramatic reversal of that scene, in the same Hall of Mirrors, was overpowering.

Whatever generosities had appeared in the opening phases of the Great War had long been exhausted. The populations of the victorious countries were acutely aware of their own losses and sufferings, and entirely regardless of the fact that the defeated had paid in the like manner. The war had arisen as a natural and inevitable consequence of the competitive nationalisms of Europe and the absence of any Federal adjustment of these competitive forces; war is the necessary logical consummation of independent sovereign nationalities living in too small an area with too powerful an armament; and if the great war had not come in the form it did it would have come in some similar form — just as it will certainly return upon a still more disastrous scale in twenty or thirty years' time if no political unification anticipates and prevents it. States organized for war will make wars as surely as hens will lay eggs, but the feeling of these distressed and war-worn countries disregarded this fact, and the whole of the defeated peoples were treated as morally and materially responsible for all the damage, as they would no doubt have treated the victor peoples had the issue of war been different. The French and English thought the Germans were to blame, the Germans thought the Russians, French and English were to blame, and only an intelligent minority thought that there was anything to blame in the fragmentary political constitution of Europe. The treaty of Versailles was intended to be exemplary and vindictive; it provided tremendous penalties for the vanquished; it sought to provide compensations for the wounded and suffering victors by imposing enormous debts upon nations already bankrupt, and its attempts to reconstitute international relations by the establishment of a League of Nations against war were manifestly insincere and inadequate.

So far as Europe was concerned it is doubtful if there would have been any attempt whatever to organize international relations for a permanent peace. The proposal of the League of Nations was brought into practical politics by the President of the United States of America, President Wilson. Its chief support was in America. So far the United States, this new modern state, had

PASSENGER AEROPLANE FLYING OVER NORTHOLT

(Photo taken from another 'plane by the Central Aerophoto Co.)

developed no distinctive ideas of international relationship beyond the Monroe Doctrine, which protected the new world from European interference. Now suddenly it was called upon for its mental contribution to the vast problem of the time. It had none. The natural disposition of the American people was towards a permanent world peace. With this however was linked a strong traditional distrust of old-world politics and a habit of isolation from old-world entanglements. The Americans had hardly begun to think out an American solution of world problems when the submarine campaign of the Germans dragged them into the war on the side of the anti-German allies. President Wilson's scheme of a League of Nations was an attempt at short notice to create a distinctively American world project. It was a sketchy, inadequate and dangerous scheme. In Europe however it was taken as a matured American point of view. The generality of mankind in 1918–19 was intensely weary of war and anxious at almost any sacrifice to erect barriers against its recurrence, but there was not a single government in the old world willing to waive one iota of its sovereign independence to attain any such end. The public utterances of President Wilson leading up to the project of a World League of Nations seemed for a time to appeal right over the heads of the governments to the peoples of the world; they were taken as expressing the ripe intentions of America, and the response was enormous. Unhappily President Wilson had to deal with governments and not with peoples; he was a man capable of tremendous flashes of vision and yet when put to the test egotistical and limited, and the great wave of enthusiasm he evoked passed and was wasted.

Says Dr. Dillon in his book, *The Peace Conference:* "Europe, when the President touched its shores, was as clay ready for the creative potter. Never before were the nations so eager to follow a Moses who would take them to the long-promised land where wars are prohibited and blockades unknown. And to their thinking he was just that great leader. In France men bowed down before him with awe and affection. Labour leaders in Paris told me that they shed tears of joy in his presence, and that their comrades would go through fire and water to help him to realize his noble schemes. To the working classes in Italy his name was a heavenly clarion at the sound of which the earth would be renewed. The Germans regarded him and his doctrine as their sheet-anchor of safety. The fearless Herr Muehlon said: 'If President Wilson were to address the Germans, and pronounce a severe sentence upon them, they would accept it with resignation and without a murmur and set to work at once.' In German-Austria his fame was that of a saviour, and the mere mention of his name brought balm to the suffering and surcease of sorrow to the afflicted. . . ."

Such were the overpowering expectations that President Wilson raised. How completely he disappointed them and how weak and futile was the League of Nations he made is too long and too distreesful a story to tell here. He exaggerated in his person our common human tragedy, he was so very great in his dreams and so incapable in his performance. America dissented from the acts of its President and would not join the League Europe accepted from him. There was a slow realization on the part of the American people that it had been rushed into something for which it was totally unprepared. There was a corresponding realization on the part of Europe that America had nothing ready to give to the old world in its extremity. Born prematurely and crippled at its birth, that League has become indeed, with its elaborate and unpractical constitution and its manifest limitations of power, a serious obstacle in the way of any effective reorganization of international relationships. The problem would be a clearer one if the League did not yet exist. Yet that world-wide blaze of enthusiasm that first welcomed the project, that readiness of men everywhere round and about the earth, of men, that is, as distinguished from governments, for a world control of war, is a thing to be recorded with emphasis in any history. Behind the short-sighted governments that divide and mismanage human affairs, a real force for world unity and world order exists and grows.

From 1918 onward the world entered upon an age of conferences. Of these the Conference at Washington called by President Harding (1921) has been the most successful and suggestive. Notable, too, is the Genoa Conference (1922) for the appearance of German and Russian delegates at its deliberations. We will not discuss this long procession of conferences and tentatives in any detail. It becomes more and more clearly manifest that a huge work of reconstruction has to be done by mankind if a crescendo of such convulsions and world massacres as that of the great war is to be averted. No such hasty improvisation as the League of Nations, no patched-up system of Conferences between this group of states and that, which change nothing with an air of settling everything, will meet the complex political needs of the new age that lies before us. A systematic development and a systematic application of the sciences of human relationship, of personal and group psychology, of financial and economic science and of education, sciences still only in their infancy, is required. Narrow and obsolete, dead and dying moral and political ideas have to be replaced by a clearer and a simpler conception of the common origins and destinies of our kind.

But if the dangers, confusions and disasters that crowd upon man in these days are enormous beyond any experience of the past, it is because science has brought him such powers as he never had

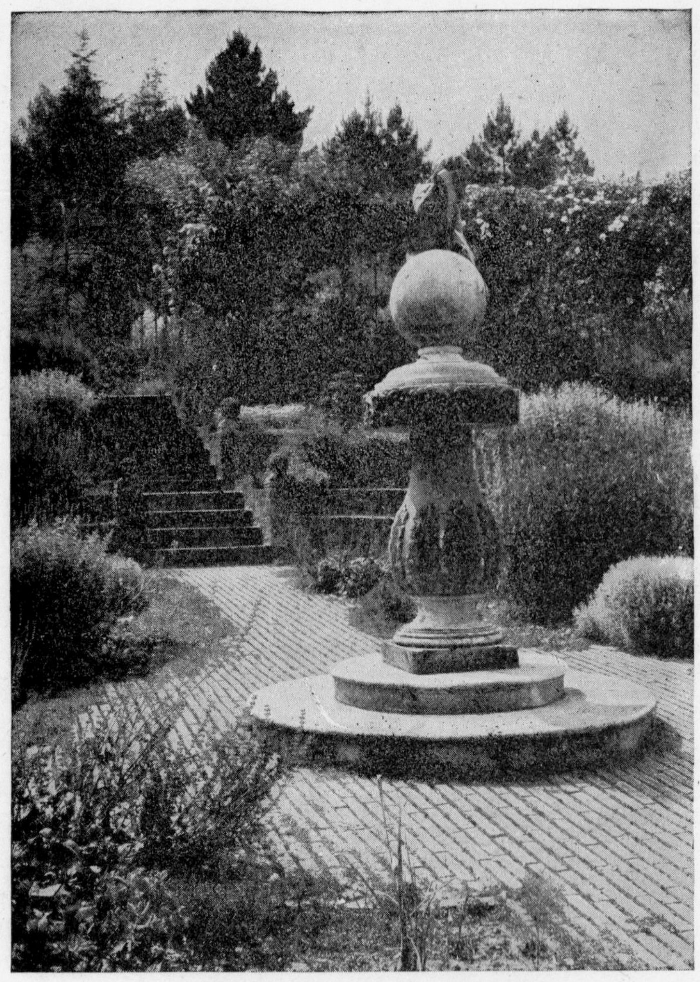

A PEACEFUL GARDEN IN ENGLAND
Given wisdom, all mankind might live in such gardens

before. And the scientific method of fearless thought, exhaustively lucid statement, and exhaustively criticized planning, which has given him these as yet uncontrollable powers, gives him also the hope of controlling these powers. Man is still only adolescent. His troubles are not the troubles of senility and exhaustion but of increasing and still undisciplined strength. When we look at all history as one process, as we have been doing in this book, when we see the steadfast upward struggle of life towards vision and control, then we see in their true proportions the hopes and dangers of the present time. As yet we are hardly in the earliest dawn of human greatness. But in the beauty of flower and sunset, in the happy and perfect movement of young animals and in the delight of ten thousand various landscapes, we have some intimations of what life can do for us, and in some few works of plastic and pictorial art, in some great music, in a few noble buildings and happy gardens, we have an intimation of what the human will can do with material possibilities. We have dreams; we have at present undisciplined but ever increasing power. Can we doubt that presently our race will more than realize our boldest imaginations, that it will achieve unity and peace, that it will live, the children of our blood and lives will live, in a world made more splendid and lovely than any palace or garden that we know, going on from strength to strength in an ever widening circle of adventure and achievement? What man has done, the little triumphs of his present state, and all this history we have told, form but the prelude to the things that man has got to do.

CHRONOLOGICAL TABLE

About the year 1000 B.C. the Aryan peoples were establishing themselves in the peninsulas of Spain, Italy and the Balkans, and they were established in North India; Cnossos was already destroyed and the spacious times of Egypt, of Thothmes III, Amenophis III and Rameses II were three or four centuries away. Weak monarchs of the XXIst Dynasty were ruling in the Nile Valley. Israel was united under her early kings; Saul or David or possibly even Solomon may have been reigning. Sargon I (2750 B.C.) of the Akkadian Sumerian Empire was a remote memory in Babylonian history, more remote than is Constantine the Great from the world of the present day. Hammurabi had been dead a thousand years. The Assyrians were already dominating the less military Babylonians. In 1100 B.C. Tiglath Pileser I had taken Babylon. But there was no permanent conquest; Assyria and Babylonia were still separate empires. In China the new Chow dynasty was flourishing. Stonehenge in England was already some hundreds of years old.

The next two centuries saw a renascence of Egypt under the XXIInd Dynasty, the splitting up of the brief little Hebrew kingdom of Solomon, the spreading of the Greeks in the Balkans, South Italy and Asia Minor, and the days of Etruscan predominance in Central Italy. We begin our list of ascertainable dates with

B.C.

800. The building of Carthage.

790. The Ethiopian conquest of Egypt (founding the XXVth Dynasty).

776. First Olympiad.

753. Rome built.

745. Tiglath Pileser III conquered Babylonia and founded the New Assyrian Empire.

722. Sargon II armed the Assyrians with iron weapons.

721. He deported the Israelites.

680. Esarhaddon took Thebes in Egypt (overthrowing the Ethiopian XXVth Dynasty).

664. Psammetichus I restored the freedom of Egypt and founded the XXVIth Dynasty (to 610).

608. Necho of Egypt defeated Josiah, king of Judah, at the battle of Megiddo.

B.C.

606. Capture of Nineveh by the Chaldeans and Medes.

Foundation of the Chaldean Empire.

604. Necho pushed to the Euphrates and was overthrown by Nebuchadnezzar II.

(Nebuchadnezzar carried off the Jews to Babylon.)

550. Cyrus the Persian succeeded Cyaxares the Mede.

Cyrus conquered Crœsus.

Buddha lived about this time.

So also did Confucius and Lao Tse.

539. Cyrus took Babylon and founded the Persian Empire.

521. Darius I, the son of Hystaspes, ruled from the Hellespont to the Indus.

His expedition to Scythia.

B.C.
490. Battle of Marathon.
480. Battles of Thermopylæ and Salamis.
479. The battles of Platea and Mycale completed the repulse of Persia.
474. Etruscan fleet destroyed by the Sicilian Greeks.
431. Peloponnesian War began (to 404).
401. Retreat of the Ten Thousand.
359. Philip became king of Macedonia.
338. Battle of Chæronia.
336. Macedonian troops crossed into Asia. Philip murdered.
334. Battle of the Granicus.
333. Battle of Issus.
331. Battle of Arbela.
330. Darius III killed.
323. Death of Alexander the Great.
321. Rise of Chandragupta in the Punjab.
The Romans completely beaten by the Samnites at the battle of the Caudine Forks.
281. Pyrrhus invaded Italy.
280. Battle of Heraclea.
279. Battle of Ausculum.
278. Gauls raided into Asia Minor and settled in Galatia.
275. Pyrrhus left Italy.
264. First Punic War. (Asoka began to reign in Behar — to 227.)
260. Battle of Mylæ.
256. Battle of Ecnomus.
246. Shi-Hwang-ti became King of Ts'in.
220. Shi-Hwang-ti became Emperor of China.
214. Great Wall of China begun.
210. Death of Shi-Hwang-ti.
202. Battle of Zama.
146. Carthage destroyed.
133. Attalus bequeathed Pergamum to Rome.
102. Marius drove back Germans.
100. Triumph of Marius. (Chinese conquering the Tarim valley.)
89. All Italians became Roman citizens.
73. The revolt of the slaves under Spartacus.
71. Defeat and end of Spartacus.

B.C.
66. Pompey led Roman troops to the Caspian and Euphrates. He encountered the Alani.
48. Julius Cæsar defeated Pompey at Pharsalos.
44. Julius Cæsar assassinated.
27. Augustus Cæsar princeps (until 14 A.D.).
4. True date of birth of Jesus of Nazareth.
A.D. Christian Era began.
14. Augustus died. Tiberius emperor.
30. Jesus of Nazareth crucified.
41. Claudius (the first emperor of the legions) made emperor by pretorian guard after murder of Caligula.
68. Suicide of Nero. (Galba, Otho, Vitellus, emperors in succession.)
69. Vespasian.
102. Pan Chau on the Caspian Sea.
117. Hadrian succeeded Trajan. Roman Empire at its greatest extent.
138. (The Indo-Scythians at this time were destroying the last traces of Hellenic rule in India.)
161. Marcus Aurelius succeeded Antoninus Pius.
164. Great plague began, and lasted to the death of M. Aurelius (180). This also devastated all Asia.
(Nearly a century of war and disorder began in the Roman Empire.)
220. End of the Han dynasty. Beginning of four hundred years of division in China.
227. Ardashir I (first Sassanid shah) put an end to Arsacid line in Persia.
242. Mani began his teaching.
247. Goths crossed Danube in a great raid.
251. Great victory of Goths. Emperor Decius killed.
260. Sapor I, the second Sassanid shah, took Antioch, captured the Emperor Valerian, and was cut up on his return from Asia

A.D.

Minor by Odenathus of Palmyra.

277. Mani crucified in Persia.

284. Diocletian became emperor.

303. Diocletian persecuted the Christians.

311. Galerius abandoned the persecution of the Christians.

312. Constantine the Great became emperor.

323. Constantine presided over the Council of Nicæa.

337. Constantine baptized on his deathbed.

361–3. Julian the Apostate attempted to substitute Mithraism for Christianity.

392. Theodosius the Great emperor of east and west.

395. Theodosius the Great died. Honorius and Arcadius redivided the empire with Stilicho and Alaric as their masters and protectors.

410. The Visigoths under Alaric captured Rome.

425. Vandals settling in south of Spain. Huns in Pannonia, Goths in Dalmatia. Visigoths and Suevi in Portugal and North Spain. English invading Britain.

439. Vandals took Carthage.

451. Attila raided Gaul and was defeated by Franks, Alemanni and Romans at Troyes.

453. Death of Attila.

455. Vandals sacked Rome.

476. Odoacer, king of a medley of Teutonic tribes, informed Constantinople that there was no emperor in the West. End of the Western Empire.

493. Theodoric, the Ostrogoth, conquered Italy and became King of Italy, but was nominally subject to Constantinople. (Gothic kings in Italy. Goths settled on special confiscated lands as a garrison.)

527. Justinian emperor.

529. Justinian closed the schools at Athens, which had flourished nearly a thousand years. Belisarius (Justinian's general) took Naples.

531. Chosroes I began to reign.

543. Great plague in Constantinople.

553. Goths expelled from Italy by Justinian.

565. Justinian died. The Lombards conquered most of North Italy (leaving Ravenna and Rome Byzantine).

570. Muhammad born.

579. Chosroes I died.
(The Lombards dominant in Italy.)

590. Plague raged in Rome. Chosroes II began to reign.

610. Heraclius began to reign.

619. Chosroes II held Egypt, Jerusalem, Damascus, and had armies on Hellespont. Tang dynasty began in China.

622. The Hegira.

627. Great Persian defeat at Nineveh by Heraclius. Tai-tsung became Emperor of China.

628. Kavadh II murdered and succeeded his father, Chosroes II.
Muhammad wrote letters to all the rulers of the earth.

629. Muhammad returned to Mecca.

632. Muhammad died. Abu Bekr Caliph.

634. Battle of the Yarmuk. Moslems took Syria. Omar second Caliph.

635. Tai-tsung received Nestorian missionaries.

637. Battle of Kadessia.

638. Jerusalem surrendered to the Caliph Omar.

642. Heraclius died.

643. Othman third Caliph.

655. Defeat of the Byzantine fleet by the Moslems.

668. The Caliph Moawija attacked Constantinople by sea.

687. Pepin of Hersthal, mayor of the palace, reunited Austrasia and Neustria.

711. Moslem army invaded Spain from Africa.

A.D.

715\. The domains of the Caliph Walid I extended from the Pyrenees to China.

717–18. Suleiman, son and successor of Walid, failed to take Constantinople.

732\. Charles Martel defeated the Moslems near Poitiers.

751\. Pepin crowned King of the French.

768\. Pepin died.

771\. Charlemagne sole king.

774\. Charlemagne conquered Lombardy.

786\. Haroun-al-Raschid Abbasid Caliph in Bagdad (to 809).

795\. Leo III became Pope (to 816).

800\. Leo crowned Charlemagne Emperor of the West.

802\. Egbert, formerly an English refugee at the court of Charlemagne, established himself as King of Wessex.

810\. Krum of Bulgaria defeated and killed the Emperor Nicephorus.

814\. Charlemagne died.

828\. Egbert became first King of England.

843\. Louis the Pious died, and the Carlovingian Empire went to pieces. Until 962 there was no regular succession of Holy Roman Emperors, though the title appeared intermittently.

850\. About this time Rurik (a Northman) became ruler of Novgorod and Kieff.

852\. Boris first Christian King of Bulgaria (to 884).

865\. The fleet of the Russians (Northmen) threatened Constantinople.

904\. Russian (Northmen) fleet off Constantinople.

912\. Rolf the Ganger established himself in Normandy.

919\. Henry the Fowler elected King of Germany.

936\. Otto I became King of Germany in succession to his father, Henry the Fowler.

A.D.

941\. Russian fleet again threatened Constantinople.

962\. Otto I, King of Germany, crowned Emperor (first Saxon Emperor) by John XII.

987\. Hugh Capet became King of France. End of the Carlovingian line of French kings.

1016\. Canute became King of England, Denmark and Norway.

1043\. Russian fleet threatened Constantinople.

1066\. Conquest of England by William, Duke of Normandy.

1071\. Revival of Islam under the Seljuk Turks. Battle of Melasgird.

1073\. Hildebrand became Pope (Gregory VII) to 1085.

1084\. Robert Guiscard, the Norman, sacked Rome.

1087–99. Urban II Pope.

1095\. Urban II at Clermont summoned the First Crusade.

1096\. Massacre of the People's Crusade.

1099\. Godfrey of Bouillon captured Jerusalem.

1147\. The Second Crusade.

1169\. Saladin Sultan of Egypt.

1176\. Frederick Barbarossa acknowledged supremacy of the Pope (Alexander III) at Venice.

1187\. Saladin captured Jerusalem.

1189\. The Third Crusade.

1198\. Innocent III Pope (to 1216). Frederick II (aged four), King of Sicily, became his ward.

1202\. The Fourth Crusade attacked the Eastern Empire.

1204\. Capture of Constantinople by the Latins.

1214\. Jengis Khan took Pekin.

1226\. St. Francis of Assisi died. (The Franciscans.)

1227\. Jengis Khan died, Khan from the Caspian to the Pacific, and was succeeded by Ogdai Khan.

1228\. Frederick II embarked upon the Sixth Crusade, and acquired Jerusalem.

1240\. Mongols destroyed Kieff. Russia tributary to the Mongols.

A.D.	
1241.	Mongol victory at Liegnitz in Silesia.
1250.	Frederick II, the last Hohenstaufen Emperor, died. German interregnum until 1273.
1251.	Mangu Khan became Great Khan. Kublai Khan governor of China.
1258.	Hulagu Khan took and destroyed Bagdad.
1260.	Kublai Khan became Great Khan.
1261.	The Greeks recaptured Constantinople from the Latins.
1273.	Rudolf of Habsburg elected Emperor. The Swiss formed their Everlasting League.
1280.	Kublai Khan founded the Yuan dynasty in China.
1292.	Death of Kublai Khan.
1293.	Roger Bacon, the prophet of experimental science, died.
1348.	The Great Plague, the Black Death.
1360.	In China the Mongol (Yuan) dynasty fell, and was succeeded by the Ming dynasty (to 1644).
1377.	Pope Gregory XI returned to Rome.
1378.	The Great Schism. Urban VI in Rome, Clement VII at Avignon.
1398.	Huss preached Wycliffism at Prague.
1414–18.	The Council of Constance. Huss burnt (1415).
1417.	The Great Schism ended.
1453.	Ottoman Turks under Muhammad II took Constantinople.
1480.	Ivan III, Grand Duke of Moscow, threw off the Mongol allegiance.
1481.	Death of the Sultan Muhammad II while preparing for the conquest of Italy.
1486.	Diaz rounded the Cape of Good Hope.
1492.	Columbus crossed the Atlantic to America.
1493.	Maximilian I became Emperor.
1498.	Vasco da Gama sailed round the Cape to India.
1499.	Switzerland became an independent republic.
1500.	Charles V born.
1509.	Henry VIII King of England.
1513.	Leo X Pope.
1515.	Francis I King of France.
1520.	Suleiman the Magnificent, Sultan (to 1566), who ruled from Bagdad to Hungary. Charles V Emperor.
1525.	Baber won the battle of Panipat, captured Delhi, and founded the Mogul Empire.
1527.	The German troops in Italy, under the Constable of Bourbon, took and pillaged Rome.
1529.	Suleiman besieged Vienna.
1530.	Charles V crowned by the Pope. Henry VIII began his quarrel with the Papacy.
1539.	The Society of Jesus founded.
1546.	Martin Luther died.
1547.	Ivan IV (the Terrible) took the title of Tsar of Russia.
1556.	Charles V abdicated. Akbar, Great Mogul (to 1605). Ignatius of Loyola died.
1558.	Death of Charles V.
1566.	Suleiman the Magnificent died.
1603.	James I King of England and Scotland.
1620.	*Mayflower* expedition founded New Plymouth. First negro slaves landed at Jamestown (Va.).
1625.	Charles I of England.
1626.	Sir Francis Bacon (Lord Verulam) died.
1643.	Louis XIV began his reign of seventy-two years.
1644.	The Manchus ended the Ming dynasty.
1648.	Treaty of Westphalia. Thereby Holland and Switzerland were recognized as free republics and Prussia became important. The treaty gave a complete victory neither to the Imperial Crown nor to the Princes.

A.D.

1648. War of the Fronde; it ended in the complete victory of the French crown.
1649. Execution of Charles I of England.
1658. Aurungzeb Great Mogul. Cromwell died.
1660. Charles II of England.
1674. Nieuw Amsterdam finally became British by treaty and was renamed New York.
1683. The last Turkish attack on Vienna defeated by John III of Poland.
1689. Peter the Great of Russia. (To 1725.)
1701. Frederick I first King of Prussia.
1707. Death of Aurungzeb. The empire of the Great Mogul disintegrated.
1713. Frederick the Great of Prussia born.
1715. Louis XV of France.
1755–63. Britain and France struggled for America and India. France in alliance with Austria and Russia against Prussia and Britain (1756–63); the Seven Years' War.
1759. The British general, Wolfe, took Quebec.
1760. George III of Britain.
1763. Peace of Paris; Canada ceded to Britain. British dominant in India.
1769. Napoleon Bonaparte born.
1774. Louis XVI began his reign.
1776. Declaration of Independence by the United States of America.
1783. Treaty of Peace between Britain and the new United States of America.
1787. The Constitutional Convention of Philadelphia set up the Federal Government of the United States. France discovered to be bankrupt.
1788. First Federal Congress of the United States at New York.
1789. The French States-General assembled. Storming of the Bastille.

A.D.

1791. Flight to Varennes.
1792. France declared war on Austria: Prussia declared war on France. Battle of Valmy. France became a republic.
1793. Louis XVI beheaded.
1794. Execution of Robespierre and end of the Jacobin republic.
1795. The Directory. Bonaparte suppressed a revolt and went to Italy as commander-in-chief.
1798. Bonaparte went to Egypt. Battle of the Nile.
1799. Bonaparte returned to France. He became First Consul with enormous powers.
1804. Bonaparte became Emperor. Francis II took the title of Emperor of Austria in 1805, and in 1806 he dropped the title of Holy Roman Emperor. So the "Holy Roman Empire" came to an end.
1806. Prussia overthrown at Jena.
1808. Napoleon made his brother Joseph King of Spain.
1810. Spanish America became republican.
1812. Napoleon's retreat from Moscow.
1814. Abdication of Napoleon. Louis XVIII.
1824. Charles X of France.
1825. Nicholas I of Russia. First railway, Stockton to Darlington.
1827. Battle of Navarino.
1829. Greece independent.
1830. A year of disturbance. Louis Philippe ousted Charles X. Belgium broke away from Holland. Leopold of Saxe-Coburg-Gotha became king of this new country, Belgium, Russian Poland revolted ineffectually.
1835. The word "socialism" first used.
1837. Queen Victoria.
1840. Queen Victoria married Prince Albert of Saxe-Coburg-Gotha.
1852. Napoleon III Emperor of the French.
1854–56. Crimean War.

A.D.

1856. Alexander II of Russia.

1861. Victor Emmanuel First King of Italy. Abraham Lincoln became President, U. S. A. The American Civil War began.

1865. Surrender of Appomattox Court House. Japan opened to the world.

1870. Napoleon III declared war against Prussia.

1871. Paris surrendered (January). The King of Prussia became "German Emperor." The Peace of Frankfort.

1878. The Treaty of Berlin. The Armed Peace of forty-six years began in western Europe.

1888. Frederick II (March), William II (June), German Emperors.

A.D.

1912. China became a republic.

1914. The Great War in Europe began.

1917. The two Russian revolutions. Establishment of the Bolshevik regime in Russia.

1918. The Armistice.

1920. First meeting of the League of Nations, from which Germany, Austria, Russia and Turkey were excluded and at which the United States was not represented.

1921. The Greeks, in complete disregard of the League of Nations, make war upon the Turks.

1922. Great defeat of the Greeks in Asia Minor by the Turks.

INDEX

INDEX

B

Index

G

H

Index

J

K

L

T